Consequences of Language

Consequences of Language

From Primary to Enhanced Intersubjectivity

N. J. Enfield and Jack Sidnell

The MIT Press
Cambridge, Massachusetts
London, England

The MIT Press would like to thank the anonymous peer reviewers who provided comments on drafts of this book. The generous work of academic experts is essential for establishing the authority and quality of our publications. We acknowledge with gratitude the contributions of these otherwise uncredited readers.

This book was set in Stone Serif and Stone Sans by Westchester Publishing Services.
Printed and bound in the United States of America.

Library of Congress Cataloging-in-Publication Data

Names: Enfield, N. J., 1966– author. | Sidnell, Jack, author.
Title: Consequences of language : from primary to enhanced intersubjectivity / N. J. Enfield and Jack Sidnell.
Description: Cambridge, Massachusetts : The MIT Press, [2022] | Includes bibliographical references and index.
Identifiers: LCCN 2022006408 (print) | LCCN 2022006409 (ebook) | ISBN 9780262544863 (paperback) | ISBN 9780262372732 (epub) | ISBN 9780262372749 (pdf)
Subjects: LCSH: Social interaction. | Intersubjectivity. | Anthropological linguistics. | Semantics.
Classification: LCC HM1111 .E536 2022 (print) | LCC HM1111 (ebook) | DDC 302—dc23/eng/20220428
LC record available at https://lccn.loc.gov/2022006408
LC ebook record available at https://lccn.loc.gov/2022006409

10 9 8 7 6 5 4 3 2 1

While we often speak of society as though it were a static structure defined by tradition, it is, in the more intimate sense, nothing of the kind, but a highly intricate network of partial or complete understandings between the members of organizational units of every degree of size and complexity, ranging from a pair of lovers or a family to a league of nations or that ever increasing portion of humanity which can be reached by the press through all its transnational ramifications. It is only apparently a static sum of social institutions; actually it is being reanimated or creatively reaffirmed from day to day by particular acts of a communicative nature which obtain among individuals participating in it.

—Edward Sapir, 1931

Contents

Preface

Imagine two points in time, deep in the history of our species.

At the first point, protohumans use a system of gestures and interjections to communicate purposefully and voluntarily but they have no strictly referential signs, not even pointing gestures. Yet they can still accomplish actions of the kind we might refer to as requesting, beckoning, and commanding, akin to what other great apes do in the wild today (Tomasello 2008; Rossano 2013).

Fast-forward to a second point, thousands of years later. The protohumans now have a fully elaborated language based on the design principles that characterize every known human language. They command a semiotic system in which a finite set of distinctive speech sounds or manual gestures combine to form a potentially infinite but still delimited set of meaningful units, including words and grammatical markers that cannot occur independently of the words to which they attach. Utterances in this language are used to refer to and predicate about any perceivable aspect of, or entity in, the world, including language itself. These protohumans can talk about particular uses of language, for example when they talk about what people mean. And they can talk about the language system, for example when they talk about what particular words mean. In and through this capacity for reference and predication, the proto-humans pursue purposive action with one another. They say things like "Come here," or "Give me some of that," or "I will do it." What's more, they say such things in response as "What for?," "I don't have enough," and "When will you do it?" There are now conversations about what was said, by whom, when, why, and with what consequence.

What has changed between these two points in time? This book proposes an answer. We do not seek to describe the evolutionary processes themselves.[1] Rather, we aim to pinpoint what is present at the second point that was absent at the first. Our concern is to understand, firstly, what modern language really is, and secondly, what are the logical and conceptual consequences of language for social life.

Our account of the consequences of language is intimately tied to the question of intersubjectivity in human social life. We argue that the uniquely human form of intersubjectivity consists of more than merely sharing knowledge or perspectives. Human intersubjectivity arises through social activity in which people pursue shared goals and where their respective contributions are observable and subject to public evaluation. With the abovementioned two points in time as coordinates, we argue that human intersubjectivity is intertwined with language in two ways. First, a primary form of intersubjectivity was necessary for language to have begun evolving in our species in the first place. Second, language then transformed the nature of our intersubjectivity, through its defining properties of inferentially articulated description, self-reflexivity, and productive grammatical flexibility. Social accountability—the bedrock of society—is grounded in this linguistically transformed and enhanced kind of intersubjectivity.

We begin in chapter 1 by explicating the concept of intersubjectivity. This is an overture to the central chapters of the book, which are grouped into two parts. Part I covers the infrastructure for human interaction, emphasizing the elements of primary intersubjectivity that serve as a set of conditions for the possibility of human language: *status* (chapter 2), *participation* (chapter 3), and *enchrony* (chapter 4). Part II emphasizes the core elements of language, which transform primary intersubjectivity and which allow language to be turned onto language itself: *reference* (chapter 5), *semantics* (chapter 6), and *grammar* (chapter 7).

Acknowledgments

Like *The Concept of Action* (Enfield and Sidnell 2017), this book is the outcome of intense collaboration. Neither of us could have written it alone.

We thank Chip Zuckerman, Chris Sinha, Michael Lempert, Chris Knight, and Luke Fleming for their valuable input on the book as we were writing it. We owe a massive debt to the hundreds of other mentors, colleagues, and students who have helped us along the way.

Enfield gratefully acknowledges the support of the Australian Research Council (through Discovery Project grant number DP170104607).

Where we have included examples from the literature, in many cases we have simplified them for readability, for example by simplifying technical interlinear glosses in the second line of the non-English examples, or by omitting non-relevant details from narrow transcriptions of conversation in English. For this reason, we urge readers to consult the original sources of examples for additional detail, especially if the examples are to be cited in further work.

We note that parts of the text incorporate revised sections of published material: Sections of chapter 1 draw on N. J. Enfield and Jack Sidnell, "Intersubjectivity Is Activity Plus Accountability," in *The Oxford Handbook of Human Symbolic Evolution*, ed. Nathalie Gontier, Andy Lock, and Chris Sinha (Oxford University Press), doi: 10.1093/oxfordhb/9780198813781.013.25; sections of chapter 4 draw on Jack Sidnell, "The Architecture of Intersubjectivity Revisited," in *The Cambridge Handbook of Linguistic Anthropology*, ed. N. J. Enfield, Paul Kockelman, and Jack Sidnell (Cambridge University Press), 364–399; sections of chapter 5 draw on Jack Sidnell and N. J. Enfield, "Deixis and the Interactional Foundations of Reference," in *The Oxford Handbook of Pragmatics*, ed. Yan Huang (Oxford University Press), 217–239; and sections of chapter 6 draw on N. J. Enfield and Jack Sidnell, "The Normative Nature of Language," in *The Normative Animal?*, ed. Neil Roughley and Kurt Bayertz (Oxford University Press), 265–278.

Many thanks to Naomie Nguyen and Gus Wheeler for their assistance during production.

We dedicate this work to our incredible kids: Nyssa and Nonnika, Sula and Ginger.

Overture

1 Intersubjectivity, Activity, Accountability

> Not every type of contact of human beings has a social character; this is rather confined to cases where the actor's behaviour is meaningfully oriented to that of others. For example, a mere collision of two cyclists may be compared to a natural event. On the other hand, their attempt to avoid hitting each other, or whatever insults, blows, or friendly discussion might follow the collision, would constitute "social action."
>
> —Max Weber (1947, 113)

Any engagement between people requires some form of intersubjectivity. As Weber suggests, a collision between strangers becomes a social exchange. Upon coming together, they adjust their orientation to the world, taking one another into account in a way that they did not before. They quickly work out (for themselves, at least) what happened, who may have been responsible, or whether anything needs to be said or done at all. And they quickly work out a solution to the problem that has arisen. One may step aside to allow the other to pass. This rudimentary kind of intersubjectivity is not limited to cases in which people collide. Rather, it is exactly what allows us to avoid walking into one another more often than we do. We take into account what others are doing, and we make detailed adjustments to our own conduct in order to get along without major incident.

Being aware of others' movements and trajectories, and heeding them in one's own actions, is an important basis for collective behavior in many species, from tracking pheromone trails in ants to schooling and flocking behavior in fish and birds. But *human* intersubjectivity obviously involves more than merely heeding others, or even being aware of what others see, want, or know. Intersubjectivity has a necessarily public, situated, and normative character. We argue that human intersubjectivity is possible only within the context of some activity in which two or more persons (or agents) participate, and in which participants are socially accountable for their participation.

Humans' complex forms of intersubjectivity allow us to develop especially complex and flexible forms of social organization.[1] We enter into arrangements with one another by which we distribute or aggregate our agency (Enfield and Kockelman 2017). This key feature of human social life has been described in different ways, each term with its own nuance: *shared intentionality*, *collective agency*, *joint commitment*, *we-intention*, and so on (Gilbert 1992; Bratman 2007; Searle 2010; Tomasello 2008; Tomasello et al. 2005; Tuomela 2002, 2007). But the basic insight is the same: two or more individuals come together to form a unit, in a passing fusion of agency, thereby overcoming the limitations that inhere in the fact that individuals have separate bodies. And with the benefits of this fusion come certain mutual rights and duties for the individuals involved.

Intersubjectivity is sometimes defined using the word "sharing," as for example in Zlatev et al.'s (2008, 1) definition of intersubjectivity as "the sharing of experiential content (e.g., feelings, perceptions, thoughts and linguistic meanings) among a plurality of subjects." For the word "sharing" to be appropriate here, it needs to be understood in a specific and strong sense. The mere fact that two people have the same beliefs or the same understandings of certain symbols or states of affairs is not sufficient for intersubjectivity (although it might qualify as "sharing" in one sense of that word). Analysts of *common ground* and related notions (e.g., Lewis 1969; Stalnaker 1999; Clark 1996) have long acknowledged that there can only be common ground if the experiential content is shared, and *this fact*—that the knowledge/experience in question is shared—is in the open (i.e., it has been made public to all parties concerned). Suppose that Kim and Pat are acquaintances but are not aware of the fact that they are both vegan. In one sense, they "share" experiential content of what it means to be vegan, insofar as this content overlaps or is largely the same. But in another, more important sense, there is no intersubjectivity in relation to this "shared" content until their commonality has been publicly registered by them, and thus brought into their common ground. Without common ground in this strong sense, their "shared" commitment to not eating animal products cannot yet serve as a basis for a range of intersubjective practices, such as the ability to induce various pragmatic inferences (e.g., by means of an eye roll upon seeing that the vegan option at a conference lunch is green salad). Or, in a more specific example, suppose that Kim and Pat both know Jo—who is Kim's cousin and Pat's work colleague—but they are not aware that they each know Jo. In this situation, Kim would refer to Jo in conversation with Pat as "my cousin," but not as "Jo." Kim would design this person reference in such a way as not to presuppose that Pat knows the person being referred to (even though, as it happens, she does know Jo). To use the "first name only" form of person reference in English, shared knowledge of the person in question is not enough. *The fact* that the experiential content is shared must also be shared (Enfield and Stivers 2007).

Our goal in this chapter is to address a key prolegomenon to any investigation of the foundations of intersubjectivity, namely the definition of the phenomenon at hand. What is intersubjectivity? It can be defined with the following equation:

intersubjectivity = activity + accountability

Contained in this equation—though not yet obvious from its wording—is our central claim, that human intersubjectivity is essentially linguistic in nature. This claim is grounded in the fact that *social accountability is both essential to full-blown intersubjectivity and impossible without language*. Language is what allows us to refer to people's words and actions in order to share attention toward those actions, characterize them in specific ways, and evaluate them publicly. In turn, the ever-present possibility of shared evaluation is the basis of social norms, which in turn are the essence of human sociality.

This claim may be taken to imply that language must have evolved before intersubjectivity could. However, we argue that a form of intersubjectivity can exist before language—and indeed it is a prerequisite for the possibility of language, in phylogeny and in ontogeny—but that language then transforms human intersubjectivity in turn. In slogan form,[2]

1. intersubjectivity makes language possible, and
2. language takes intersubjectivity and transforms it.

Language and human intersubjectivity are co-constituting. Language has intersubjective foundations, all the way down to its core properties. But intersubjectivity and language do not have a symmetrical relationship. On the one hand, an infrastructure of intersubjectivity is foundational for language (Tomasello 2008). Language cannot exist without it. The infrastructure is a precondition for the development of language in the life span, and surely also in the evolution of our species. On the other hand, the human form of intersubjectivity that we observe today acquires unique properties because of language. Nonlinguistic animals have their own forms of intersubjectivity,[3] but language allows for a ramping up, a jump, a significant qualitative elaboration of intersubjectivity, in which defining properties of language are key. These defining properties include (1) the possibility of segmentable/thematizable inferentially articulated description, (2) self-reflexivity of communication (by which linguistic behavior can be used to refer to linguistic behavior), and (3) the productive grammatical resources for building unique moves in interaction, enabling complete flexibility of language-mediated action.

In the remainder of this chapter we elaborate these proposals through a consideration of three mundane forms of activity. In each we identify a form of intersubjectivity that is inherent in (both presupposed by and required for) the organization of the activity itself. Furthermore, in each of these activities we can see the role, from relatively minor to deeply significant, that language plays in making that organization possible.

1.1 A Simple System of Human Activity: The Two-Person Saw

The images in figure 1.1 are stills from a video recording of two men (monks from a temple in lowland Laos) sawing planks using a two-person saw. With this type of saw, two people work together to cut a plank lengthwise from the topmost layer of a large chunk or log of wood. In the foreground at the right of the image in figure 1.1, the top part of the plank can be seen beginning to lift up as it separates from the main block.

Smooth coordination here entails complete complementarity of action. As the two panels of figure 1.1 show, when one man has pulled the saw all the way back, the other man must necessarily move all the way forward, and vice versa.

Sawing like this implies special kinds of organization, both for the who and when of turn-taking and for the what of action sequencing.[4] While we readily speak of "taking turns" in many situations, such as when playing board games, it is better to think in terms of opportunities to participate in social interaction. Any coordinated joint activity, from conversation to

Figure 1.1
Two-person saw in operation. (Doune Ian Village, Vientiane, Laos.)
Still image from video by N. J. Enfield.

ballroom dancing, from roadwork to open-heart surgery, requires some way of organizing and managing the contributions of the various people who are involved. As Schegloff (2000b, 1) puts it: "The orderly distribution of opportunities to participate in social interaction is one of the most fundamental preconditions for viable social organization." In the form of social interaction we are considering here, the system leaves little to chance. The possibilities are tightly constrained by the artifactual structure at hand and the activity in which it is employed.

To the extent that each man knows how the saw operates, he *subprehends* what the other man is about to do. We might say "anticipate," but this would imply a conscious expectation of what is to come. The term *subprehend* captures more accurately what we want to say. To subprehend something is

Figure 1.1
(continued)

not necessarily to consciously or actively expect it but rather to not be surprised or disposed to sanction it when it does happen (and, correspondingly, to be surprised or disposed to sanction it when something happens that is not subprehended; Enfield 2013, 23; Enfield and Sidnell 2017a, 65). This concept was introduced by Kockelman (2007b) under the label *commitment*. The idea is akin to the sociological–anthropological notion of *habitus* (Bourdieu 1977; Hanks 2005), but here we emphasize the regimenting force of the tyranny of accountability for social action (Enfield and Sidnell 2017a, Enfield and Sidnell 2019).

The result is a microcosm of smooth sociality. Within this highly structured environment, two individuals operate as if they were one. The human-oriented design of the artifact at hand (*lùaj1 ñaj1,* "large saw" in Lao)[5] is an important part of what makes this possible. But there are other equally important requirements. One concerns the extent to which each person slots themselves into a tightly constrained role. As long as they are engaged in the activity of operating the *lùaj1 ñaj1*, the men pictured in figure 1.1 effectively suspend their individual identities. For the purposes of the activity, they become nonindividuals. But because they have separate bodies, the partners must work to maintain their shared agency. They must both focus their attention on what both parties are doing. Each contributes what is required under the assumption that the other will do the same. As a result, the participants can quite reliably subprehend or anticipate what the other one is going to do, and when.

There may be a theoretically perfect complementarity between the contributions of the two participants, but in practice there can be asymmetries. For instance, in one of the sawing sequences we observed, the saw was being operated by two monks very different in age and capacity. One was old and frail, the other young and able. The younger monk provided most of the physical power, pushing and pulling the saw through the wood at each stroke, while the older monk, using just one hand rather than two, merely helped by guiding the saw along the correct path.

Despite the simple, constrained nature of the sawing activity, participants sometimes encounter trouble, which must be resolved there and then if the activity is to stay on track. As the video recordings show, there are occasional snags of execution, such as when the saw catches a knot in the wood and momentarily refuses to budge. When this happens, the activity of sawing is suspended while one or both participants try to free the tool. Occasionally, participants stop to extract the saw completely, for example to adjust or sharpen it. Both participants are committed to overcoming these snags to move the activity forward, though this of course can be flouted for effect. If one monk were to jokingly pull when it was his turn to push, this might elicit laughter. But it may just as likely elicit annoyance, particularly if repeated. It would disrupt the activity and would also inevitably condition expectation of future action (will he do that again?) and thus erode trust between partners.

The smooth sociality of working with the two-person saw involves sustained mutual engagement and joint commitment to a shared goal. By agreeing to participate, both participants must maintain attentional focus and must continue to contribute in just the way required by the activity.

These are transient commitments. They are like the commitments involved in working together to assemble a piece of furniture (Clark 1996) or taking a walk together (Gilbert 1990). We call them transient not because they are unimportant but because they are highly localized and largely confined to the scope of some delimited activity. In our sawing example, once the desired planks have been cut, the men's mutual focus and commitments are no longer required or maintained.

There are other, more enduring forms of commitment. Consider the fact that the sawing in this scene is done by monks. In Laos, a village man can be a monk for anything from a few hours to weeks, months, or years. Whatever the duration, the status of monkhood entails—indeed is constituted by—certain rights, duties, and goals that implicate the entire community. With this commitment, individuals give up or suspend their individuality to a degree, as strikingly emblematized in the uniform clothing (or "habit") in which the monks are dressed. Here the participants' commitments are scaled at a different level, but the result is strikingly similar to what happens when two people work together to saw planks of wood. Taking on the assigned status requires limiting one's own individuality and thereby making one's own actions more subprehensible, and more publicly accountable. This in turn facilitates intersubjectivity. In subsequent chapters we will show that this basic operation by which agency is narrowed and effectively guided plays a key role in all forms of social interaction.

The activity of sawing can be understood as a form of exchange: A → B, B → A, and so forth. The "good" being exchanged is immediately reciprocated and the entire process is rigidly structured by the physical affordances of the saw itself. But this simple form of exchange is itself embedded in other, more complex forms. For instance, B is giving his labor to A and will expect, in return, to be compensated (though not perhaps by A's gift of his labor). And the two, working as a unit, are sawing this wood for a larger collective project (temple construction), which is itself a move in a series of exchanges that reaches into the future just as it responds to what has come before. The momentary action captured in our screenshots in figure 1.1 is constituted by these different dimensions of shared understanding.

Intersubjectivity is thus in one way instantaneous but in another way time-independent, flowing across and drawing upon an indefinite set of understandings, experiences, role relationships, and so on. Notably, in the case of the sawing monks, this is all unsaid. There is no discussion, in the local context of this scene at least, of who they are to one another and why they are doing this work. These are matters of common ground. In other cases, however, and in a wide range of ways, such commitments are brought into the here and now of the communicative encounter, making them relevant to and oriented toward aspects of the intersubjective context.

Let us here reemphasize the role of language. The accountability required for creating and maintaining intersubjectivity in joint activity would be impossible without language. Holding somebody accountable for having failed to meet their commitment within a joint activity—the potential for "rebuke" that Margaret Gilbert points to in her 1990 analysis of joint action (see below)—is only possible if their behavior can be thematized and characterized.[6] For that, we need a way of describing what was done (or not done). Language gives us this.

We shall see how this works as we now move on to examine a kind of joint activity that is a step up in complexity from the two-person saw but is nevertheless simpler and more constrained in its structure than free-flowing conversation.

1.2 A Step Up in Complexity: Reed-Mat Weaving in Pairs

In lowland rural Laos, villagers engage in a seasonal practice of weaving reed mats using a *Cyperacea* species of marsh reed known locally as *phùù3* (a sedge, related to papyrus), which villagers gather in the rainy season (July–September).

The reeds are collected, dried, and dyed different colors prior to being woven into large mats (called *saat5* in Lao). These mats—of dimensions roughly the same as a large tabletop—serve a range of indoor and outdoor domestic uses.

Like sawing with a *lùaj1 ñaj1* "large saw," weaving reed mats is a two-person task. It is usually done by women at a large, flat loom placed on the floor in the family living area. Figure 1.2 shows a weaving session in operation.[7]

In figure 1.2, the mother has her left hand on a large wooden block that is angled forward, so as to make a crosswise opening between even and odd warp threads. The daughter has a long stick in her right hand, and with her left hand she has selected a reed of the appropriate color (as determined by the pattern they are weaving). Each reed is finely tapered, with a thick end (the base of the plant) and a thin end (the tip of the plant). The daughter is using

Figure 1.2
A mother-daughter pair weaving reed mats. (Doune Ian Village, Vientiane, Laos.) Still image from video by N. J. Enfield.

her left hand to hook one end of the reed over the tip of the stick (when placing each reed, she must alternate between thick end first and thin end first, so that the mat will be structured evenly), in preparation to run the reed through the held-apart warp yarns and across the width of the loom, such that it will come to rest with its two ends poking out at each side of the structure.

Figure 1.3 shows the daughter launching the stick-with-reed through the opening that the mother creates by pushing the wooden block forward.

As figure 1.4 shows, once the reed is through, resting with its two ends protruding from each side of the loom structure (at the mother's left and right, respectively), the daughter extracts the feeder stick from the structure, leaving the reed in place, so that the mother can pull the wooden block back hard toward herself, bringing the newly inserted reed tightly into the structure of the mat.

Figure 1.3
Daughter launches stick with reed hooked over the end. (Doune Ian Village, Vientiane, Laos.)
Still image from video by N. J. Enfield.

Once a reed is in place, with its base and tip poking out at either side of the mat structure, the mother ties off one end of the reed—the thick end—securing it to the growing structure of the mat. This tying-off action is done at the thick end for each reed, which alternates from left to right. Figure 1.5 shows the mother tying off the end of the reed to her left (then, with the next reed, she will tie it off to her right, and so on).

In each phase, the participants are subprehending the next phase. In figure 1.5, as the mother is tying the end of the weft to her left, she is simultaneously pressing back on the block with her right hand so as to create an opening in the weft for her daughter's next move (alternating pressing forward and back each round so as to ensure that the weft goes over, and then under, each warp thread); and by selecting a reed weft in figure 1.4

Figure 1.4
Mother pulls wooden block back toward herself, wedging the reed into place. (Doune Ian Village, Vientiane, Laos.)
Still image from video by N. J. Enfield.

and hooking it over her stick, the daughter has subprehended (i.e., effectively anticipated) the moment illustrated in figure 1.5, so that she is ready to launch the stick and weft through the structure as soon as this becomes possible. The two participants are subprehending each other's moves.

Figure 1.6 maps out the interlocking and open-endedly reiterating phases of action that each individual must carry out, where the actions of one of the participants depend on, and complement, those of the other.

We video-recorded a number of these weaving sessions and examined 254 reed-insertion cycles, by four pairs of weavers. We found variation in how quickly and fluently different partners can work. The average times for each pair are shown in Table 1.1.

The mother-in-law/daughter-in-law pair are nearly three times faster at this task than the husband/wife pair. The husband/wife pair (pictured in figure 1.7) worked together for a period when the husband was asked to

Figure 1.5
Mother ties off end of reed weft to her left. (Doune Ian Village, Vientiane, Laos.) Still image from video by N. J. Enfield.

fill in for his daughter-in-law, who had to nurse her baby. The husband has seen the women of his family weaving in his living room thousands of times, but had evidently never paid close attention to the process.

Table 1.1

Weaver pair	Average time per cycle
mother-in-law/daughter-in-law	5.3 s (n = 72)
neighbor/neighbor	9.9 s (n = 86)
mother/daughter	11.1 s (n = 68)
husband/wife	14.6 s (n = 28)

A feature of the type of interaction that caused this pair's slow progress is the need for linguistic commentary to make explicit what is normally tacit. Here is a fragment from the interaction:

Feeder	Blocker
find new weft of color x; hook weft over feeding-stick, thick end first	[hold warp yarns apart with block forward]
<insert stick-with-reed through between warp yarns>; [remove stick, leaving loose weft protruding from warp at both sides]	wait
find new weft of color x; hook over stick, thin end first	<pull block in>; push block out; [hold warp yarns apart with block backward]
<insert stick and loose weft between warp yarns>; [remove stick, leaving loose weft protruding from warp at both sides]	tie off jammed weft at thick end
find new weft of color x; hook over stick, thick end first	<pull block in>; push block out; [hold warp yarns apart with block forward]
<insert stick and loose weft between warp yarns>; [remove stick, leaving loose weft protruding from warp at both sides]	tie off jammed weft at thick end
find new weft of color x; hook over stick, thin end first	<pull block in>; push block out; [hold warp yarns apart with block backward]
<insert stick and loose weft between warp yarns>	tie off jammed weft at thick end
etc.	etc.

Figure 1.6

The interlocking and interdependent actions of the "feeder" and "blocker" roles in two-person reed-mat weaving in lowland Laos. Time runs from top to bottom. The successful completion of each action [in square brackets] is a necessary prerequisite for the subsequent action <in angle brackets>.

Figure 1.7
Husband/wife pair at the loom. The husband is a novice, filling in temporarily while his daughter-in-law is nursing her baby. (Doune Ian Village, Vientiane, Laos.) Still image from video by N. J. Enfield.

(1) (Lao, data ref: INTCN-020714 10.05)

01	H	—	[pulls back on block, then presses it to face up, incorrectly (it should be pressed facing down on this cycle)]
02	W	—	[watching H, preparing next weft]
03		—	*nguam5* *long2* **on.front** **down** "[Press the block] down on its front."
04	H	—	[presses block to face down, then waits (but should be tying off)]
05	W	—	[points to site for tie-off]
06		—	*phaj2* *mèè4* **tie-off** **do-it** "Tie [it] off!"

07 — [waits, holding weft ready, visually monitoring H's tie-off]

08 H — [gazing at tie-off point, ties off]

09 — *cang4-sii4 mèèn1 bòò3*

like-this right right?

"Like this, right?"

10 W — [gazing at tie-off point]

11 — *qee4, kiaw1 maa2 cang-san4 lèq1, qee4*

yeah, coil come like-that just yeah

"Yeah, coil it around just like that, yeah."

12 — *cùù1 na mèè4—bòò1 cak2 cùù1 vaa3*

"Remember it, can't you remember it?!"

13 — [inserts weft]

14 H — [laughs]

15 — [ties off]

16 — *qoo4 man2 ñaak4 nòq1*

"Oh, it's difficult."

In line 1, the husband is pulling back on the top of the block to open the warp yarn for the next reed, but he should be pressing it forward for that cycle. In lines 2 and 3, the wife is simultaneously (1) preparing her own next action, (2) monitoring the husband's action, and (3) issuing a linguistic command for him to do the required action (which is the opposite of what he is currently doing). He makes the correction in line 4 but then does not go on to carry out the required next action, namely tying off the thick end of the weft reed that has just been inserted. His wife is monitoring this, and in lines 5 and 6 she simultaneously finger-points to where the weft needs tying-off and verbally commands him to tie it off. She then waits (in line 7) and watches him work. In lines 8 and 9, now with his own gaze focused on the tie-off point where his hands are busy, the husband verbally invites his wife to confirm that he is correctly doing what is required. In lines 10 and 11, she responds, while gazing at his hands doing the action, with a rather elaborate confirmation. The confirmation not only includes the simple interjection of confirmation *qee4* "yeah," but also a linguistic description of what he is doing as being correct—"coil it around just like that"—thus giving him explicit evidence that it is correct, by using language to specify on record just *how* it is correct (i.e., in his way of coiling it rather than in any of the infinite other aspects of what he's done). Finally, there is some reparative banter in lines 12–16, with elements of mock-berating by the wife, and self-effacement by the husband (note his self-directed laughter in line 14; cf. Jefferson 1984).

1.3 Language as a Regulating Tool

Transcript (1) illustrates language in its role as a tool for coordinating joint activity (see Malinowski 1935; Goodwin 2000; Clark 1996). In (1), words are being used to draw explicit attention to things that the actors are doing, or need to do, in order to publicly thematize those things, invite evaluation of them, draw attention to problems, and elicit clarification and advice, all in the service of coordinating joint action toward a shared goal. The two individuals in the interaction have had next to no experience weaving together, and most importantly, one of the parties does not share the knowledge and skill needed to do what elite weavers do. Language is an indispensable tool for dealing with his lack of expertise. It is true that in (1), the interjected talk about the activity itself is a partial cause of the slow pace of this pair's interaction. But while explicit talk slows things down when compared to the activity of elite weaving pairs, without the possibility of using language to coordinate their behavior, this pair would be even slower.

Faster, experienced pairs still occasionally need language as a tool for dealing with snags or interruptions. For example, an issue that comes up for even the most experienced weaving pairs is the question of which color reed needs to be inserted next. This is handled linguistically, but the exchanges are quite stripped-back (in a way reminiscent of Wittgenstein's "slab/block/beam/pillar" game; see Wittgenstein 1953, §I.2ff). Here is an example, from a recording of an experienced weaving pair (two neighbours). Prior to selecting the next reed, the weft feeder asks for confirmation that red is indeed the color now needed (only the English translation is provided here):

(2)

01 A: "Red?"

02 B: "Yeah."

03 A: [selects red]

While the weft feeder is conveying that she is not sure which color is needed next, she is putting forward a specific proposal, that it is red. She could have said "Which color?" but instead chooses to be as specific as possible.[8]

An example from a mother/daughter pair shows a case of linguistically mediated other-correction. The daughter has selected a red reed when a green one was required. The mother, who is monitoring throughout, uses words to draw attention to the problem—in a form of other-correction of bodily conduct—and the daughter makes the correction without any verbal uptake (only English translation provided here):

(3)

01 A: [selects red]

02 B: "First, another green." (as instruction to use green instead)

03 A: [puts down red, gets green]

In line 2, language is used to draw attention to a problem and to define what the problem is. In line 3, language can be elided, as the daughter's action not only makes the required correction but shows the mother, who is monitoring the situation, that the required correction is being made. Mutual monitoring is crucial here (Goodwin C. 1979, 1981; Goodwin M. H. 1980).

While the last two examples show only minor disruptions to the flow of otherwise-fluent weaving interaction, perhaps the most important feature of the fastest and smoothest cycles is that elite weavers do not talk at all. Or if they do talk, their talk is not about their weaving. When members of a weaver pair have common knowledge, mutual monitoring, shared goals and subgoals, reciprocities of perspective, and complete trust that the other will know what needs to be done, and will do it at the right moment, then the two individuals can effectively act as a single agent with no need for the talk that would otherwise be necessary to thematize, characterize, and compensate for failings of individual or joint performance. The smoothness of an interaction can thus look deceptively easy. These weaving scenes are a significant step up in complexity and unpredictability from the monks' activity of sawing timber boards, where heavy constraints of the activity maximize the chances of a seamless, single-agent, machine-like organization of cooperation.

1.4 Accountability and Norms

We have emphasized that accountability is essential to cooperative interaction, and to human intersubjectivity in general. There is a *tyranny of accountability* in human affairs (Enfield and Sidnell 2017a): an ever-present possibility of being noticed, praised, blamed, questioned, called out, and judged.[9] We act knowing that our actions will be observed and, to a significant extent, knowing how they will be regarded.[10] As sociologists since Goffman (1961) and Garfinkel (1967) have argued, this is the basis of social norms, the bedrock of human sociality. Note that norms are not the same as laws (Kockelman 2006b).[11] Norms provide points of reference for accountability, which in turn give people coordinates for regulating their own social behavior.[12] A norm is not written down and thus cannot be pointed to in the same way a law can. In this way, norms are necessarily indeterminate. Also, unlike a law, a norm is not undergirded by the legitimate threat of force within a specific institutional context. Nevertheless, norms do a highly effective job of shaping our social behavior.[13]

Three key properties of norms help to explain why this is so. If a pattern of behavior is supported by a norm, it will have the following characteristics:

1. ***Subliminal*** *(the behavior's presence is unnoticed).* When the behavior occurs, no particular attention is drawn to it at a meta-level; nobody is within their rights to sanction the behavior or express surprise that the behavior occurred. People *subprehend* the behavior (i.e., without necessarily or actively expecting it, they are not surprised when it happens) (Enfield 2013, 23 and passim).
2. ***Abliminal*** *(the behavior's absence is noticed).*[14] When the norm is violated—when the expected behavior is missing and/or something else occurs in its place—this can be noticed and reported or remarked upon.
3. ***Inference-vulnerable*** *(absence of, or deviation from, the behavior generates inferences).* When the norm is violated—when the expected behavior is missing and/or something else occurs in its place—people will assume that the person who produced this violation did so for a reason; people will try to infer what the reason was, and they will look for meaning there.

See how this applies to our example of the two-person sawing operation. If a sawing partner is following the norms of the activity—maintaining a timely rhythm of alternation complementing the other's pushing and pulling—then their actions will, firstly, be subliminal. That is, their actions will not be remarked upon, nor even really noticed, as they will be entirely expected. When things are going normally, partners simply saw away without comment. Conformity is invisible. Second, if someone does depart from the norm-defined behavior—for example if a partner does nothing when they should be pulling the saw through—then this *will* be noticed and will likely be remarked upon. And third, in turn, observers of any such departure will make inferences that rationalize the departure. For example, we might surmise that the person is tired and having a rest, or that they are messing with us, as may befit our social relationship.

Each of these properties of norms relates directly to accountability. Another way to define the subliminal nature of normative behavior is to say that the conforming behavior does not attract sanction, surprise, or demands for explanation. And if people do draw attention to the behavior or try to seek accounts for it, then they themselves are accountable for this (i.e., if people are following norms, then others have no right to demand an account).[15] Norm-sanctioning behavior is itself sanctionable, thus giving rise to an infinitely extended prospective–retrospective web of normative action. The abliminal nature of nonnormative behavior means that a person who produces such behavior should not be surprised when others hold them accountable, with expressions of surprise and/or sanction. And the inference-generating nature of nonnormative behavior will determine the likely *content* of acceptable accounts for the behavior: thus, we may say we pause because we are tired. The ubiquity, omnirelevance, and simultaneous operation of these properties of norms (and their web-like interconnections to other norms) gives rise to what we call the tyranny of accountability.

When Margaret Gilbert (1990) defined going for a walk together as a kind of joint action, accountability was at the core. She described it in terms of the possibility of "rebuke." Gilbert noted that participants may rebuke each other for not doing their part, as when one of two walkers starts moving ahead: "You are going to have to slow down! I can't keep up with you" (Gilbert 1990, 3). This is what we mean by a tyranny of accountability with reference to norms. In our sawing and weaving examples, as in all forms of intersubjective activity, participants are always a move away from being rebuked, or indeed from having the right—or duty—to rebuke another. People avoid rebuke by carrying out their duty as defined by the norms of the activity. Indeed, the desire to avoid being the object of attention or rebuke may be enough to provide people with an incentive to conform: that is, people conform not because they want to be like others but because they want to avoid the stigma that comes from being seen as different (Goffman 1963b). The awareness of, and orientation to, norms in any joint activity provides the framework needed for intersubjectivity.

Let us say more about the concept of accountability, as it is central to most of the ideas in this book. *Accountability* is of course an English word, and is loaded with cultural meaning, and is as contested and negotiable as any such term. In a simple sense of the word, being accountable is being "expected or required to account for one's actions; answerable."[16] More technically, accountability could be defined as "the obligation of an individual or organization to account for its activities, accept responsibility for them, and to disclose the results in a transparent manner."[17] Or, in further detail, it is "the obligation imposed by law [on a] person for keeping accurate record of property, documents, or funds. The person having this obligation may or may not have actual possession of the property, documents, or funds. Accountability is concerned primarily with records, while responsibility is concerned primarily with custody, care, and safekeeping."[18]

As these definitions suggest, the concept of accountability has multiple elements, and these rely on language in multiple ways. Accountability requires "record-keeping," which in turn requires that our actions not only be monitored but linguistically defined or described in some way. It requires that we thematize the relevant piece of behavior or situation, and that we will do this not by merely pointing to it but also by characterizing it in some way, usually using words. And it requires that we orient to the need to provide reasons, something that again cannot be done without language.

As an illustration of the function of reasons in regulating human interaction, consider their role in the issuing of requests. Baranova and Dingemanse (2016, 30) note that "reasons serve to make requests more readily understandable and increase the ultimate likelihood of compliance." Sometimes, a bare request is issued and it is simply complied with, as in the following case:

Request with no reason:

A. Can you give me a spoon?

B. [passes spoon]

On other occasions, though, the request is not met with compliance, but with some signal of a problem, such as a repair initiator (e.g., "What?"), which will often then elicit not a redoing of the original request but the giving of *a reason* why the request is being made:

Request → Problem → Reason:

A. Can you give me a spoon [holding a spoon]

B. What?

A. This one's dirty.

B. [passes spoon]

In their study of requesting behavior in everyday home/village interactions in Russian, Baranova and Dingemanse (2016, 644) show that while a majority of requests (63 percent of their cases) are not accompanied by a reason that accounts for why the request is being made, nearly a third of the time (32 percent) a reason for the request will be provided by the requester before compliance occurs (e.g., as illustrated in the "This one's dirty" example, shown above). People are so strongly oriented to the giving of reasons that they will even respond to "reasons" that are entirely devoid of content, so long as they are presented as reasons. Langer, Blank, and Chanowitz (1978) conducted an experiment in which people who were busy using a photocopy machine were approached by someone asking to cut in and copy five pages. When no reason was given, people complied around two-thirds of the time. But when a reason was given, they complied more than 90 percent of the time. Crucially, this high rate of compliance occurred both for "real" reasons (e.g., "because I'm in a rush") and for empty reasons, which added no content at all to the original request to make copies (i.e., "because I have to make copies").

Reason-giving and reason-demanding play an essential role in the architecture of human intersubjectivity. And they are linguistic in nature. To provide reasons for an action is a public act, which entails settling on a description of that action that is fit for a purpose, and committing to any consequences that those reasons may have. Reason-giving illustrates the essentially linguistic nature of the tyranny of accountability.

1.5 Accountability and Agency

To understand how accountability may operate within the context of intersubjective activity, we need to examine it in terms of the *agency* of the people involved.

Agency is not defined as a simple value on a scale of more versus less but is defined as the combination of an actor's *flexibility* and *accountability* (Kockelman 2007b). Component loci of agency can be defined in terms of control, composition, and subprehension (see Kockelman 2005, 2007b; Enfield 2013, 91–93), defined as follows:

1. An agent **controls** a perceptible event to the extent that they can determine the timing and placement of the execution of a piece of behavior, and as a corollary, that they can attend to the timing and placement of another's behavior.
2. An agent **composes** a meaningful action to the extent that they select and execute a specific behavior that should stand for something, and as a corollary, that they can interpret others' actions as standing for things.
3. An agent **subprehends** to the extent that they are not surprised by what happens next as defined by the conventions of the activity, as calibrated to the conventionalized rights and duties of participants.

Accountability can be located relative to any or all of these elements of agency. In terms of control, each party is accountable for their timing and placement in executing the relevant behavior. For example, in our weaving cases, if one partner delays the execution of their required action—say, inserting a reed—for no obvious reason, this will cause surprise and possible sanction, and the other partner might wonder why the delay is occurring. In terms of composition, each party is accountable for the selection and execution of the specific behavior chosen. This would apply to, among other things, the selection of a specific color of reed. We saw that when a partner selects the wrong color, they are held accountable. And in terms of subprehension, each party is surprised or disposed to sanction behaviors that depart from the norms at hand, and if they fail to express such surprise or sanction where appropriate, this itself may be accountable ("Why didn't you tell me I put in the wrong color?!"). Similarly, if they express surprise or sanction where none is warranted, this too is accountable (see Raymond and Sidnell 2019).

1.6 Language Use as an Intersubjective Activity

We have contrasted two highly constrained collaborative activities observed in Lao village life: the two-person saw and the two-person loom. While these differ in complexity, they are extremely simple compared to the collaborative activity known as conversation. Conversation lacks the mediating artifactual structure needed for sawing or weaving (i.e., a piece of physical technology such as a saw or a loom), but this does not mean that the activity is simpler. Far from it. The tools for conversation are the bodies of the participants, and these provide many more degrees of freedom than a saw or a loom for what can be done, how, when, and by whom in the course of the activity. The seemingly unbounded flexibility of talk in interaction is enabled by an infrastructure for human interaction (Schegloff 2006; Levinson 2006, 2019; Enfield and Sidnell 2014; Sidnell 2014; Enfield 2017). Here are some of its core elements:

- **Participation**: a system for entering, maintaining, exiting, and otherwise navigating membership in the activity, grounded in mutual monitoring and joint commitment (Goodwin and Goodwin 2004; Sidnell 2009c)
- **Turn-taking**: a system for regulating who gets to speak and when (Sacks, Schegloff, and Jefferson 1974); may involve real-time projection of timing and content of turn-endings
- **Action sequencing**: a system for regulating what actions are done, by whom, and in what order (Schegloff 2007a); may require interpretation via intention-ascription (Enfield and Sidnell 2017a; Sidnell and Enfield 2017)
- **Repair**: a system for identifying and resolving problems in speaking, hearing, and understanding (Schegloff, Jefferson, and Sacks 1977; Hayashi, Raymond, and Sidnell 2013; Schegloff 2013; Dingemanse et al. 2015)
- **Generativity/Productivity**: the property of the sign system used in interaction that allows people to generate entirely novel meaningful structures (Hockett 1960; Chomsky 1957; Croft 2001; Adger 2019)
- **Self-referentiality**: the property of the sign system used in interaction that allows people to use the system to refer to itself (Hockett 1960; Jakobson 1960, 1985, 1990a, 1990b; Lucy 1993)

We can isolate some of the elements of these parts of the infrastructure for social interaction, in order to directly compare the three forms of collaboration we have considered so far. These elements are featured in table 1.2.[19]

Practical activities like sawing and weaving result in forms of intersubjectivity that are sufficient for the execution of that activity, but the required degrees of shared understanding are relatively minimal. If you and I are

Table 1.2
Elements of the infrastructure for interaction, compared across three activity types

	Sawing	Weaving	Conversation
Common ground for activity assumed?	yes	yes	yes
Dyadic mutual monitoring needed?	yes	yes	yes
Joint commitment needed?	yes	yes	yes
Language available for explicit regulation of the activity?	yes	yes	yes
Content of each turn known in advance?	yes	yes	no
Actor and order of each turn known in advance?	yes	yes	no
Relative no. of turn-units by each participant known in advance?	yes	yes	no
Public visibility of formulation of next turn?	extensive	extensive	minimal
Roles symmetrical?	yes	no	no
Choices to be made at each move?	none	limited	unlimited
Complexity of mediating artifact?	low (saw)	medium (loom)	high (body)
Move composition	binary	limited	free
Move control	dependent	mixed	independent

engaged in the activity, you know what I am doing with my body but you do not necessarily know, for instance (and I don't necessarily convey), how I feel about each push, pull, insertion, or tie-off. More importantly, I don't display a detailed or nuanced understanding of your previous move in each response (beyond my recognition that your action/turn is complete). In these practical activities, the physical behavior executed with each turn in itself constitutes the action that is being performed. By contrast, in conversation—an activity that is, by definition, mediated by the conventionalized semiotic structures of language—each move requires both semantic decoding and pragmatic interpretation (among much else), drawing in part on the conventions of meaning in the signs selected, and in part on inferences deriving from the fact that a conversational response is always a choice from a technically infinite number of possible responses. In sawing and weaving, if there is selection of a next action, it does not discriminate between, or otherwise reveal, a selector's alternative qualitative understandings of a preceding move. It is merely generated by the structures and procedures at hand. The loom uses the weavers as much as the weavers use the loom.

The activity of talk is more complex, both quantitatively, as language has so many degrees of freedom, and qualitatively, given the unique semiotic properties of language. In conversational structure—comprising such things as turn construction, turn-taking, action sequencing, action formation, and repair—agents have elaborate possibilities for control and composition. The high degree of flexibility means that much more can be conveyed in each move. We have noted that in activities such as two-person sawing and weaving, each "turn" is also a physical, practical action. But linguistic exchanges require us to deal with turn-taking and action formation independently of each other. Consider this example (from Schegloff 1988c):

(2)

01 Mom: Do you know who's going to that meeting?
02 Russ: Who?
03 Mom: I don't know.
04 Russ: Oh, probably Mrs. McOwen and
05 probably Mrs. Cadry and some of the teachers, and the counselors.

Mom's utterance in line 1 has clear lexical content and grammatical structure, but it is ambiguous in terms of the action it is meant to perform. In one of its possible functions, it is an information question: Mom is asking who's going to be at the meeting. In that case, an appropriate response from Russ in line 2 would be to give the information requested—that is, for Russ to tell Mom who's going to be there. In another of its possible functions, the question in line 1 is preliminary to telling the addressee who's going to be there (with a meaning something like "Guess who's going to be there"). In that case, an appropriate response from Russ in line 2 would be to say "Who?" (with a falling intonation) as a go-ahead, which would then be followed up by Mom in supplying the promised information (see Terasaki [1976] 2004; Schegloff 1988c).

In the example given here, Russ demonstrates, by starting to speak (in line 2), that he has perfectly well understood that Mom's turn at talk in line 1 has come to completion, and that the moment is right for him to begin his own turn (he also shows that whatever Mom was doing with her talk, she has selected him as its primary recipient). But we can also surmise from line 2 that Russ has understood line 1 as a preliminary rather than the question that Mom meant it to be. To summarize what has happened here, in line 2, Russ has got the turn-taking right but the action wrong. This possibility does not arise when the moves at hand constitute practical actions in themselves, as for example in two-person sawing or weaving.

The man who pushes the saw forward is not doing anything else *by* pushing it forward. He is just pushing it forward.[20] His success in doing this is wholly consummated by physical mechanisms and outcomes. He needs the cooperation of the other man, but the direct functional outcome of pushing the saw is not defined by, nor affected by, the other man's *interpretation of* what he has done. Two-person sawing requires intersubjectivity for its operation, because the two people involved must effectively act as one. But it does not require intersubjectivity for its effects to be realized. This is where language is different. Such intersubjective interpretation is precisely what linguistic action depends on. The kind of intersubjectivity that language requires and creates adds a distinct layer on top of the form of intersubjectivity that is needed for practical collaborative action. This is because the linguistic actions themselves—we sometimes call them speech acts—cannot be done unless there is a shared understanding that they are being done. In the sawing case, those effects are realized by the brute force of serrated steel upon timber. With language, actions are only fully consummated by the appropriate (i.e., subprehended, unsurprising, nonsanctionable) recognition or attribution of an action value to the behavior (Enfield and Sidnell 2017a).

Because interpretation is necessary for language to have its effects, structures of conversational interaction require detailed and nuanced forms of displayed, shared understandings so that participants get evidence of what each move has effectively achieved. The unique referential, characterizing potential of the semantic and grammatical structures of language leave a wide berth for interpretation at every step. Just think of how heavily we rely on radically underdetermined deictic expressions such as *this*, *here*, and *us*. This does not mean that every usage of a vague or ambiguous term is understood—or must be understood—exactly as it was intended. But there is a threshold by which people's understandings should not deviate intolerably from what was intended (Enfield 2015). Because social interaction in the enchronic frame proceeds in a move-by-move manner, there are constant opportunities for participants to encounter evidence showing when understanding was not as intended. And language is the tool for dealing with it when such evidence arises, as when we initiate repair with expressions such as "What?," "Who?," and "Huh?"

But as we've shown, as long as things are going smoothly in a cooperative activity—whether it is conversation or a practical activity such as two-person weaving—we will tend not to talk about or refer to that activity unless we have to (Enfield 2006; see Rossi 2014 for a case study). Here is the reason. To talk about what we are doing betrays a need to inform, and thus a lack of common ground, a failure on the part of those involved to subprehend the next action or phase of behavior in the activity at hand and, therefore, a sign of something lacking in the degree and quality of common ground in the personal relationship between participants (Enfield 2013, chap. 13). The more we engage in "second-track" work—work that explicitly regulates the activity at hand (Goffman 1963a; Kendon 1990; Clark 1996)—the less on track we show ourselves to be. One might even say that the entire reason for talking at all is to compensate for the fact that we do not share the same body. It is precisely when individuals successfully merge as a single corporate person—as elite mat weavers do—that we least need explicit forms of communication.

Many scholars have emphasized the often-unspoken nature of cultural convention. Goodwin (2013, 19) says that elite practitioners "have mastered the infrastructure of skilled practice," noting the link to the concept of habitus (Bourdieu 1977). Habitus is, by definition, unarticulated, but it is foregrounded precisely when violations of norms occur, in accordance with the abliminal nature of norms—thus, habitus is essentially normative. Habitus draws on the human capacity for subprehension, our ability to effectively anticipate what is to come, and what is to be done, in ways that do not surface propositionally (i.e., through language or some equivalent or proxy) unless there is reason to express surprise or to invoke accountability, whether negatively in the form of critique or blame, or positively in the form of praise. This point applies to talk about talk itself. As Goffman pointed out, the "main-line track" of conversation is coordinated by a "directional track", in which there is "a stream of signs which is itself excluded from the content of the activity but which serves as a means of regulating it, bounding, articulating and qualifying its various components and phases" (1974, 210). In talk, the directional track is where we find much talk-*about*-talk, from repair to other practices of accountability that operate via language's unique capacity to thematize and characterize actions that are done through language itself.

The preference for avoiding explicit reference to a cooperative activity as it proceeds does not contradict our claim that language is indispensable for intersubjectivity. It is not that people need to talk to achieve intersubjectivity. Our sawing and mat-weaving examples show this. But while elite mat weavers may not need to talk about mat-weaving while they are doing it, language still underlies the activity's existence. The activity of mat-weaving would not be possible in the first place without language having laid the agreed-upon groundwork for establishing the expectations and requirements that define this collaborative activity (Searle 2010). It would not have become possible had language not mediated participants' socialization into the activity. And words are always waiting in the wings, for those times when a trajectory of activity diverges from expectation.

1.7 Conclusion

Language and human intersubjectivity are co-constituting. Language presupposes intersubjectivity, but once language develops, it supercharges that intersubjectivity, ramping up the possibilities for shared understanding and effecting an essentially qualitative shift. Another way of putting it is that intersubjectivity is activity dependent. The first and most obvious sense in which this is true is in the fact that there can be no intersubjectivity without some form of mutual engagement, even if this is as seemingly simple as one canine smelling the urine of another canine on a tree. Intersubjectivity is a feature of engagement with others, including but not limited to conspecifics. Primordially, intersubjectivity is a relation between two parties (e.g., an infant and its mother) but this basic situation is, in human interaction, spun out in innumerable ways and eventually, ontogenetically and presumably phylogenetically, includes joint attention to a third party (Tomasello 2003, 2008; Kockelman 2005).

We have explored some subtler aspects of the relation between intersubjectivity and activity, proposing that specific human activities require and essentially constitute particular forms and degrees of intersubjectivity. As the activity becomes more complex, along any number of dimensions, so too do the understandings that participants necessarily share.

The endpoint is spontaneously occurring human conversation, in which shared understandings can be incredibly detailed (with tone and stance and coloring in addition to the central notion of what a person "means by" what they say), and constantly updated *en passant*. In the enchronic time frame, interaction is essentially built up out of moves, from turns at talk, with each one layering on what has come before, in this way revealing the actors' understandings, which can then become the target of repair or other remedial action.

Compared with sawing and even weaving, language-mediated interaction is assembled out of what appears to be an infinite range of possibilities. (Later in this book, we will look more closely at the generative grammatical resources of language.) But it is not merely the complexity of conversation that defines human intersubjectivity. It is that linguistic communication is reflexive. We are continuously communicating about the very act of communication: the current act, past acts, and potential future acts. The possibilities for communicating *about* sawing *by* sawing, or *about* weaving *by* weaving, are, in comparison, almost nil, at least not without bringing language in. Without the capacity to thematize and characterize our own actions, we would lack the backdrop of social accountability that forms the basis of human relationships and institutions.

*

This completes our overture to the core business of the book. We now turn to part I, which is focused mostly on the infrastructure for social interaction. We begin, in chapter 2, with the concept of status.

I Prerequisites for Language

2 Status and the Tyranny of Accountability

The concept of intersubjectivity necessarily introduces concepts of personhood, agency, and the self. Does intersubjectivity presuppose the existence of individual, subjective selves? Or can individual selves only emerge from intersubjective relations? We argue that there is no need to choose between these two possibilities, because each notion—subjectivity and intersubjectivity—really depends on the other (Benveniste 1966). Rather than treating this as a hopeless circularity, we embrace a concept of the two-way relationship between the agency of individuals and the intersubjectivity created and experienced by groups of two or more people. That concept is *status*.

2.1 Status Defined

A person's status—in the technical sense we will soon define—determines their rights and duties to signify and interpret in various ways. It is a way of defining agents in relation to other agents. We have expectations of people defined in terms of the status that they claim or that we assume they have. For instance, a general practitioner is obliged to know basic human anatomy as well as the names and symptomatology of various ailments. A rugby fan is expected to know the rules of rugby (among much else relating to the game). And a presidential candidate is expected to be up to speed with current affairs. When expectations are not met, a person's assumed or claimed status may disintegrate.

Consider the example of US presidential candidate Gary Johnson when he was interviewed by CNN journalist Mike Barnicle in September 2016:

(1)

01	CNN:	What would you do, if you were elected—about Aleppo.
02	Johnson:	About . . .

03	CNN:	Aleppo.
04	Johnson:	And what is Aleppo?
05	CNN:	You're kidding.
06	Johnson:	No.
07	CNN:	Aleppo is in Syria. It's . . . the epicenter of the refugee crisis.
08	Johnson:	Okay, got it. Got it, well—with regard to Syria, uhm, I do think that it's a mess . . .

After Johnson fails to recognize the reference to Aleppo—a name and place that was heavily referenced in political news at the time—he tries to recover by turning the topic to Syria more generally. But the damage was done. *The New York Times* wrote that Johnson "revealed a surprising lack of foreign policy knowledge," which for many critics meant that Johnson had thereby "disqualified himself from the presidency."[1] That is, Johnson's failure was taken as a sign that he did not meet shared expectations of the status of "US presidential candidate" (or indeed, "US president").[2] Notably, various explanations for what happened appealed to other kinds of statuses: as a "marijuana user," Johnson was being forgetful; as an "isolationist," he was revealing a lack of interest in foreign affairs; as a "human," he was only being human by making a mistake. The example throws into high relief the accountabilities associated with a status, and the invocation of alternative statuses in reasoning about departures from expectation.

Ideas about status lie at the heart of social, political and legal theory. Any discussion of rights or duties, whether in Plato, Hegel, Hobbes, Rousseau or Paine, implicates notions of status. Our interest is more narrowly defined. We aim to indicate the ways in which human intersubjectivity—as a species-specific infrastructure for mutual understanding—is inherently tied to notions of status. The word *status* here is used in a technical sense and not in the everyday senses of "rank" (*high status*) or "state" (*What's your current status?*). The anthropologist Ralph Linton (1936) distinguished between *status* and *role* (or *rôle*, as he put it), with status referring to one's position in a social system (e.g., "son-in-law") and role referring to the behavior expected of a person in a certain position. And Linton (1936, 188) further defined status as "simply a collection of rights and duties."[3] Similarly, Kockelman (2006a, 51) defines status as a bundle of "commitments and entitlements to signify and interpret in particular ways."

We define status in relation to agency and role with reference to the semiotic process described by Charles Sanders Peirce (see Peirce 1933; Kockelman 2005; Enfield 2013). The three parts of the semiotic process are *sign*, *object*, and *interpretant*. A sign is anything that can be perceived (e.g., a dark raincloud approaching). An object is anything that a sign may be taken to stand for (e.g., the dark raincloud may indicate that *rain is imminent*). An interpretant is any response to the sign that makes sense insofar as it orients to a connection between the sign and the object (e.g., grabbing an umbrella on the way out the door, or bringing in washing from the clothesline, are responses to seeing a dark raincloud approaching that make sense insofar as someone takes the raincloud to stand for imminent rain).[4]

Semiotically, statuses are constructed on the basis of evidence from the signs and interpretants that orient to those statuses. We focus here on statuses as human kinds, that is, as inherently reflexive forms of categorization. This means that a person who is categorized in a certain way will find that their thoughts and behaviors are guided by that very categorization.[5] This insight has been developed, for example, by Hacking (1986, following Foucault 1978, 1980) through the idea of the looping effects of human kinds in "making up people," and by Taylor (1985) through the idea of humans as "self-interpreting animals." A status provides a person with a means to understand their own or others' conduct. In the sense we are interested in here, status always exhibits a special kind of reflexive quality.

To serve as a rule and so to guide the production and interpretation of human conduct, a status must be public. To paraphrase Wittgenstein (1953), if a status were not public, we would not be able to tell the difference between acting in accordance with the status and just seeming to act in accordance with it.[6] Some kind of external criterion of assessment is definitional of the very notion of a rule and thus of status.

Does this imply that statuses must be named in order to serve as human kinds? While many statuses do have conventional names, the point is not whether a status is named per se but whether it can be *publicly recognized* as a type and so become linked to a normative set of rights and duties. Consider, for instance, the status of "current speaker" in a conversation. As we discuss in later chapters, this is a status with a clear and robust set of rights and duties—for example, the right and duty to complete the current turn at talk, and the duty of others not to interrupt the current speaker—but it is not typically invoked by use of a status name per se. Upon being interrupted one does not say "I am the current speaker," but rather "I'm still speaking," "I'm not done yet," or "Don't interrupt me." This may be because, typically, the rights and duties of a person *qua* speaker are invoked in close proximity to the relevant conduct, which obviates the need for a name. This can be compared with statuses such as *mom* and *dad*, which are frequently invoked at some distance from the conduct that is understood in relation to them. As an example, in September 2019 it was reported that a couple had been "arrested after spending US$120K mistakenly added to their account." When asked to comment on the matter by a news station, the father of one of the two people charged responded: "I don't even know what's going on," adding "I'm just the dad,"[7] thereby challenging the implicit assumption that he, as a parent, should have some insight into or knowledge about the situation.

When a status is labeled, this brings the concept into public discourse, such that we may coordinate around the idea when we use the label for it, and in turn, the idea is more likely to enter into our practices of reason-giving. Indeed, having a conventional word for the status makes community members accountable for using the concept to predict and explain people's behavior.

In the 1960s and 70s, professional anthropologists began to use the status label *big man* to refer to individuals in Pacific societies who showed a certain kind of leadership quality and political status earned through skill in economic endeavor (Lindstrom 1981, 903; see Sahlins 1963; Finney 1973; Burridge 1975; Hawkes 1978; Keesing 1978; Sillitoe 1978). The term was applied to a diverse range of statuses in a diverse array of cultures and did not in fact correspond to a native term used in any of the relevant linguistic communities. If this was truly a status in the cultures being investigated, it was not a named one. In his analysis of this piece of technical labeling, Lindstrom argued that as soon as we label something we have observed, we become constrained by that label: "the first time we match word to concept we use language; the second time, repeated, language begins to use us" (1981, 900).

We interpret Lindstrom's paradoxical remark—that as soon as we use a status label, the label "begins to use us"—as follows. Because a word labels a *category*, it strips away differences between individual instances. The categories of language lead us to overlook those differences. This is good when it means we are not distracted by irrelevant differences between members of a set. But sometimes those details are important. If we look closely at reality, we may not find sufficient evidence that a unified category really exists, but because the word in question is now in circulation and acts as a label, our interpretations will take it into account. As Lindstrom says, it is not so much that we are using the word, but the word is using us.

Statuses, as labeled by words, have a range of values and orientations. Some statuses are more contestable than others. With *police officer*, either you are or you aren't one. The same goes for the word *pregnant*. With *hipster* or *punk rocker*, people might disagree as to whether a person qualifies as inhabiting that status, or who genuinely fits into that category, and it might be impossible to adjudicate. Another distinction is how much information a status gives, and so how powerful it may be as a generator or predictor of behavior. Values expressed by status terms can range widely. Some status terms are neutral, lacking any strong inherent connotations or valence (e.g., *plumber, uncle, citizen*). Others carry inherent evaluative valence, such as positive terms like *warrior* or negative terms like *terrorist*. Another parameter is the degree of symmetry of a status term's application. With some terms, all speakers of a language have the same rights to apply the term to a person, assuming that the person meets the criteria. There are no constraints on who can call whom (including oneself) a *plumber* other than that the person should meet the relevant criteria. Other terms are inherently asymmetrical. For instance, Harvey Sacks (1979, 11–12) noted that some categories "are owned by a group other than those to whom they apply," while others are actively controlled by group members. His example involves the categories of *hotrodders* and *teenagers*. The latter, he says, is controlled by adults; it is "a category that adults administer" and "what's known about 'teenagers' is enforced by adults" (1979, 11). In contrast, "what is known about hotrodders—what they do with their cars, how they look, how they behave—these are things that hotrodders can enforce on each other and defend against nonmembers". With the status of hotrodders then, it's up to the hotrodders to decide "whether somebody is a member" and "what that membership takes," and it's hotrodders who "do the sanctioning" (12).

As this example suggests, status terms often convey an evaluative stance, and their application may be resisted on that basis. For example, because the term *terrorist* carries a particularly negative connotation—and because there can be highly negative consequences of being classified as one—we would predict that people are more likely to label others with this term than themselves. The opposite would be true with positive terms such as *freedom fighter*. Taking these examples together, we may conclude that statuses are typically perspectival (e.g., insider–outsider, member–nonmember, incumbent–nonincumbent), that they are often tied to other statuses with which they form a relatively coherent set (mother–father–daughter, hotrodder–cop), and that they provide a rich basis for inference about those who use them (revealing age, gender, ethnic affiliation, political position, temperament, etc.).

The public quality of a status varies from cases in which the status is almost universally known, recognized, and supported by extensive metasemiotic discourse and institutional infrastructure (e.g., by a courtroom judge or a customs officer) to those in which it is fleeting, ephemeral, and relevant only for those two people engaged in some joint activity (e.g., taking a walk together, sawing a plank together). This has two interrelated implications.

First, a status, like any other semiotic phenomenon, operates over some social domain. The domain may be defined in terms of the actual extent over which knowledge of the relevant signs is distributed. In some cases, this may be correlated with knowledge of a relevant language. In other cases, the social domain of a status is restricted by occupation, education, training, political positioning, and so on. The irreducibly social dimension to status—like any other semiotic phenomenon—must at some point be materially realized and thus travel along networks of communication. As we discuss in chapter 4, all signs (including the signs that point to statuses) must at some point run through the semiotic bottleneck in enchronic occasions of language use. A status cannot exist only as a type but must be signified within the stream of conduct in a material form that can be interpreted and responded to, and can provide a basis for inference. This realization of signs is also a mechanism by which such signs are distributed and continue to circulate in a population.[8]

Second, a status is never fully consistent across the domain within which it operates. A given status may entail different rights and duties within different communities and some statuses may exhibit more internal consistency than others. Much of this depends, of course, on the degree to which the discourse surrounding a status is elaborated. For instance, in English-speaking North American communities, there is a pervasive and perpetually self-renewing discourse about the status "mother," which can be found in everyday conversation, self-help books, online forums, newspaper editorials, and so on. The discourse surrounding "cousin" is, in comparison, hardly developed at all.

There is a reason why statuses are widely contestable and difficult to pin down. It is that a status is an *object* in a semiotic process (Kockelman 2006b, 2007a). An object is not directly observed but is only ever pointed to by signs and their interpretants. With signs and interpretants as data, we hypothesize as to what the objects of those signs and interpretants must be (Enfield 2015; Enfield and Zuckerman in press). Objects are thus "forever deferred" (to use Derrida's apt words, but perhaps not in the sense he may have meant). A person's understanding of a sign's object need not be exact or definitive, but it does need to be good enough to allow them to make sense of the sign-interpretant relations we generate or observe.

While *agency* refers to a person's general *capacity* to signify and interpret in various ways, status refers specifically to our *rights and duties* to signify and interpret in particular ways (Kockelman 2006b, 51).[9] By incorporating the notions of rights and duties, the notion of status is inherently social-relational. Rights aren't just things you can do, but things you can do without expecting others to express surprise or sanction. Duties aren't just things you must do, but things that, if you *don't* do them, you *can* expect surprise or sanction from other people. Rights and duties also often imply a form of social reciprocity: if I have a right to do something, you have no right to stop me (Searle 2010).

Status implies social relationships not only through the inherently social nature of rights and duties, but also through the fact that statuses tend to have reciprocal orientations. Perhaps the clearest examples are kin relations. For instance, the status of *mother* invokes the status of *child*. Norms around a person's status as "mother" will specify not just rights and duties but rights and duties in relation *to the person's child*. Most obviously, they have the duty to protect and nourish the child. And in certain circumstances they have the right, and sometimes the duty, to physically restrain the child (but not other people's children). They have the right to bestow a name on the child (but not on other people's children). And in turn, norms around the status of *child* will specify rights and duties in relation to a mother.[10]

Before we flesh out the idea of status in light of our interest in intersubjectivity, we want to emphasize two crucial points.

The first is that, just as with the elements of the semiotic process we have noted already, there is no definitive or single status that a given person inhabits. A person has any number of statuses at once, but these statuses are not all relevant or explanatory at the same time. A person may be a mother and pregnant and an adult and a musician and a foodie and a US citizen. That they have a right to vote in US federal elections follows from their statuses as a US citizen and an adult, while the other statuses are irrelevant. To say that a certain status plays a role in a given context is a claim, and as such that claim is contestable and may need to be supported with evidence.

The second point is that status, as a bundle of rights and duties, is *generative* and *predictive*. To the extent that we understand the collections of rights and duties that a status entails, this means that, as actors, when we inhabit a given status we can generate a set of behaviors that would not surprise others, and when *others* inhabit a given status we can anticipate the kinds of behavior that *they* will produce. In this way, perhaps the most important function of status is to *reduce uncertainty* for agents, both in signifying and interpreting, in specific contexts of social interaction. And to the extent that people's models of statuses in their community are accurately inter-calibrated, they can conduct their interactions with maximal confidence and minimal turbulence. This calibration is the essence of intersubjectivity.

A simple and pervasive illustration is the predictive power of external signs of status. As a sign of status, a police officer's uniform is "maximally public and minimally ambiguous" (Kockelman 2006b, 53, 2007a, 156). The uniform doesn't just tell us that the person is a police officer. It allows us to assume a wide range of obligations (e.g., they should give us assistance if we ask) and entitlements (e.g., they may detain us or fine us if we commit a crime or misdemeanor) as well as abilities (e.g., they will be trained in using firearms, and in administering first aid). This enables us to make all manner of quite reliable predictions concerning the person wearing the uniform, predictions that we are confident others would also make, and that the uniformed person can subprehend, reducing uncertainty all around.[11] Less formal emblems than uniforms also allow us to predict a lot about a person, though in less defined ways.[12]

2.2 Institutional Infrastructures of Status

Statuses are subject to complex social embedding. Some statuses are inherently tied to (though not fully determined by) biological processes of maturation. These include *infant, child, teenager, adult,* and *senior citizen.*[13] Others are tied to institutional structures that are organized around explicit rules or laws, a constitution or a charter. Others are tied to forms of exchange that may also be supported by a legal framework. Some are tied to institutions of education and training that are uniquely authorized to certify or grant degrees to those who graduate from them. Still other statuses, such as current speaker in a conversation, dealer in a game of bridge, or guest and host at a wake, are tied to activities and are thus less enduring. Despite substantial differences, these types of status share a common core. They all presuppose a jointly ratified institutional infrastructure, whether that lasts for seconds or centuries. There is a cline of formality. Minimally, these infrastructures coordinate the metasemiotic discourse through which a status as type is constituted. Maximally, they may provide explicit means for regulation, enforcement, and sanction.

This variation in formality also occurs in other domains of social reality. An example is currency. The official national currency of a nation-state is supported by a massive legally grounded infrastructure of laws, mints, banks, funds, and markets, as well as by the innumerable everyday transactions in which people repeatedly orient to the values associated with paper banknotes and other stand-ins. Such values endure for as long as the institutional structures that support them do, which could be years, decades, or even centuries. Compare this to a situation in which friends get together to play poker and assign to the chips different values according to their color. These values last for the length of the game. Or consider children who, after Halloween, trade (barter) their candy. One of these candies—a particularly common and less desirable type—can be treated as something akin to small change, an addition to one side of an exchange to even it out.

Similarly, there is variation in the degree of institutionalization and infrastructural support that statuses have. At the more formal end, consider the case of Vietnamese *người có công với cách mạng*: "meritorious people" or "people who have contributed significantly to the Revolution." People with this official status designation are entitled to forms of preferential treatment and this is enshrined in law. People who participated in the revolution before January 1, 1945 are entitled to

- monthly benefits, monthly allowances;
- medical insurance, convalescence and recovery, and the provision of essential aids and orthopedic tools;

- access to the *People's Newspaper* and other appropriate cultural and spiritual activities; and
- support from the State in the form of a housing improvement allowance, based on the merits and circumstances of each person.

When such revolutionary activists die, those who organize the funeral are entitled to an allowance and burial fee and the father, mother, spouse, and/or children are entitled to a survivorship allowance.

Formalized statuses often need to be socially marked in unmistakable ways, so that members of a community can converge around intersubjective understandings of the inferences that signs of status enable. When those signs are maximally public and minimally ambiguous, uncertainty is at a minimum. This applies, for example, in the case of the monks we described in chapter 1. Like police, they wear a uniform: saffron robes, shaven heads and eyebrows, and various paraphernalia. Members of Lao-speaking culture and society recognize that certain expectations (or more accurately, subprehensions) about the behavior of monks are conventional in their society. Among many readily cited injunctions are that monks should

- be male
- not drink alcohol
- not eat after midday
- keep their hair and eyebrows shaven
- always wear monks' robes and garments
- partake in routine village services and events
- partake in daily dawn processions and accept food offerings from villagers
- avoid travel outside their local area during "Buddhist Lent" (rainy season)
- keep distance from women: avoid touching them, avoid passing things to them directly, avoid being alone with them
- use special pronouns (e.g., for "you" and "me") when addressing laypeople

When we say that the two men we discussed in chapter 1 each have the status of "monk," we do not mean that they necessarily follow these injunctions (though they probably do, for the most part). We mean that (1) if they *do* follow these injunctions, people *will not* be surprised or disposed to sanction them, and (2) if they *do not* follow these injunctions, people *will* be surprised or disposed to sanction them. Also, crucially, as monks the two men cannot reasonably deny that they know their status, what it entails, and the average person's understanding of it. They can subprehend people's dispositions toward them. So, for example, if a monk is seen eating food after midday, he should not be surprised at or disposed to sanction anybody who is surprised at or disposed to sanction *him* for the transgression.

The status of monk, with its defining entitlements and obligations, becomes a possible point of reference for evaluating behavior. A monk who flirts with women, who eats after midday, or who fails to turn up to temple events may be sanctioned, and may be evaluated (e.g., in village gossip) as a bad monk. And the status of monk may be invoked in sanctioning a nonmonk for acting like a monk in one or another of the key respects listed above. So, if a regular villager won't accept an invitation to womanize or drink with his friends, they may sanction him with a statement like "Are you a monk or something?!"

Now of course as individuals, the two men in figure 1.1 sawing boards in monks' robes also inhabit other statuses. They are not only "monks" to other people, they are many other things at the same time: brothers, friends, sawing partners, and more. When we say that they are monks, or when we focus on our subprehension of how monks act, we background those other statuses. But those other statuses are still there. They can be invoked and foregrounded where relevant. For example, in the scene we described in the opening section of the book, we focused on the expectations that the men have of each other as partners in a collaborative activity of sawing planks from a large log of wood. The fact that they also happen to be monks is independent of, and essentially irrelevant to, their rights and duties as sawing partners. Consider some injunctions that apply to these individuals not as monks but as co-sawyers. They each have the following obligations:

- pushing the saw forward while the other is pulling back, and vice versa
- paying attention to where the saw is entering the wood, and maintaining a straight line
- continuing with the task until the goal is completed
- providing a reason if they have a need to pause or abandon the task

Again, they may not always follow these injunctions. The injunctions capture the norms that account for people's tendency to express surprise or sanction at others' behavior, and to reason about their behavior (including giving reasons for, and making inferences about, behavior that departs from what is anticipated or subprehended). The injunctions also guide the men's behavior insofar as the men are motivated to conform with the norms, if only to avoid attracting surprise or sanction.

These men's incumbency in a status has a causal basis in a socially shared agreement among people that they do indeed have these statuses. In this sense, status is a piece of what Searle (2010) describes as institutional reality. Status is created by *declarations*, which are overt linguistic agreements (or proxies for them). Minimally, this agreement is a reciprocal one between the two people involved. In the case of the two men's status as co-sawyers, the agreement they made may have been simple and highly informal, such as a proposal-uptake sequence (A: "Let's go saw." B: "OK.") or some equivalent (including nonverbal equivalents such as pointing and nodding). As Gilbert (1990) described for the joint activity of going for a walk together, once the parties have entered into the activity together, they have become mutually committed, and are liable to be rebuked or to receive some other sanction if they fail to maintain the same walking pace as the other person or unilaterally abandon the activity.

While the agreement that commits two people to doing some sawing may be established in an immediate and informal way, the agreement that commits a man to the obligations and entitlements of being a monk is, by contrast, a highly formal, ritualized one. The formal rite of passage that a person must go through in order to become a monk has a name in Lao: *buat5*. This elaborate set of initiation procedures must be carried out by specified actors (senior monks with formal authority), using circumscribed ritual language, under circumscribed conditions (at a specific location in the temple, at a specific time of day, on a specific day, etc.; see McDaniel 2008). The emblematic and focal event of the initiation ritual is the shaving of the monk's head and eyebrows and the donning of saffron robes, both of which result in the unmistakeable "maximally public and minimally ambiguous" (Kockelman 2006b, 53) signs of the new status. There is also a named process through which somebody *exits* the monkhood, termed *sik2*. This rite of passage (which consists of a simple declaration and is not associated with formal ritual) causes the person to no longer have the entitlements or obligations appropriate to a monk.[14]

A man must continue to wear the monk's formal attire if he is to continue to hold the status of monk. This doesn't mean that anyone wearing the uniform is a monk. (An actor in a TV show doesn't become a monk when he puts on robes for the role.) The ritual must have been performed under appropriate felicity conditions if the declaration is going to have its true effect (Austin 1962). If an unauthorized person were to perform the ritual, or if it were performed at the wrong place or time, the person won't have been successfully initiated, and therefore won't be a monk, though we wouldn't necessarily know that just by looking at him. As with a person wearing a police uniform, we simply assume that the person is legitimately wearing it. In the case of the police uniform, if this turned out to be untrue, this would be a punishable offense. In New York, for example, impersonating a police officer—a way of claiming the rights associated with that status—is a matter of criminal fraud and is punishable by up to four years in prison.[15]

Now, suppose a man has been ordained and rightfully signals his status as monk and enjoys the entitlements. He is undoubtedly a monk but may be regarded as a poor example of one if he does not follow the precepts as strictly as he should. If he were to have a glass of beer, or to eat something after midday, this would not mean that he would no longer be a monk in any official sense (unless repeated transgressions led to him being excommunicated) but he would gain a reputation as a lax monk or something similar. There is a tradition of joking among Lao speakers about exactly these types of behavioral transgressions. In a recording of two Lao men in conversation, they say that after midday, monks "don't eat rice" (the Lao idiom meaning "having a meal"), then they laugh and joke that monks might have fruit or noodles instead. There are also more risqué jokes about monks asking young laywomen to lean further forward—thus exposing their cleavage—while placing rice in the monks' rice bowls during morning almsgiving.

The use of modifiers such as "actual" or "real" are frequent in contestation of social statuses (on the use of "real," see Austin 1962, 62–77). They can imply that would-be incumbents are not true or genuine members of a category. Ethnic labels often attract discourse of this kind. In the Nam Noi valley of upland central Laos, two neighboring communities who both refer to themselves with the ethnonym *Kri* speak closely related dialects of the same language, have similar cultural practices, and are in close, frequent contact (Zuckerman and Enfield 2022). One of the two groups, living in the more isolated upstream reaches of the valley, has come to be known as *Kri Tàn*, which literally means "real Kri" or "true Kri." People in the *Kri Tàn* community are more conservative in their language and cultural practices. The more progressive downstream-dwelling group are still labeled as

Kri (or to be more specific, *Kri Phòòngq*, an ethnic label), but they are not "real Kri." In the context of the shared culture and society of all Kri speakers, both upstream and downstream, this real/not-real distinction does not carry a strong evaluative import. There is nothing inferior about not being "real Kri," and if anything it's the other way around, as the "real Kri" are regarded as "lower" in class terms. But the "real Kri" distinction does have consequences for the statuses defined by membership in one or the other community. For example, an obligation that comes with the status of being a "real Kri" woman is to observe special taboos during menstruation. These taboos are not observed by Kri Phòòngq women (see Zuckerman and Enfield 2022 for more information).

Status—as a kind of semiotic object—is by its nature never directly inspectable, and because many statuses are not marked as publicly and unambiguously as with the monks' saffron robes, with status there is often ambiguity. When can we be sure that a person really inhabits a particular status? How do we decide? That ambiguity can become charged as a site of contestation. For example, in rape trials, defending lawyers have often pursued the strategy of building the case that the woman involved is of "low moral character" and/or is somehow responsible for causing her attacker to infer that she was inviting or consenting to sexual advances. This is nothing other than an attempt to establish a particular status in order to secure preferential outcomes for the defendant. Such attempts can define people's legal rights and duties and may mean the difference between imprisonment and freedom.

A striking case of the legal consequences of an ambiguity in status is that of Omar Khadr, a Canadian citizen whose family was claimed to be associated with Al Qaeda (Shephard 2008; Liss 2012). In July 2002, Khadr (aged fifteen at the time) was staying in an Al Qaeda safe house in Khost near the eastern border of Afghanistan. A US Special Forces unit attacked the compound. After a four-hour firefight and bombardment, Khadr was captured. It was later alleged that, during the fighting, Khadr had thrown a grenade that killed US Delta Force soldier Sergeant Christopher Speer. Khadr was transferred to the US military facility at Guantanamo Bay, where he was held for ten years and subjected to torture and inhumane treatment by US forces (Liss 2012, 96–99). In protracted legal arguments after Khadr's return to Canada in 2012, questions arose around the ambiguities of his status during his capture and subsequent detention. The answers to these questions would determine whether Khadr had been entitled to (and deprived of) certain protections while in custody. Specifically, there is a legal distinction between the statuses of *combatant* and *civilian* in international humanitarian law (Liss 2012, 110ff) according to which, in international armed conflicts, a combatant is entitled to *combatant's privilege* (e.g., they may kill enemy combatants and not risk domestic prosecution for murder) and to be treated as a *prisoner of war* upon capture (with protections under the Geneva Conventions). At first, Khadr was detained on the basis that the situation was not defined as an international armed conflict and that therefore he could be (a) charged with murder and (b) detained and interrogated without the protection of prisoner of war status (as was the case with other detainees at locations including Guantanamo Bay in Cuba and Abu Ghraib in Iraq).[16] Because the United States took Khadr to be affiliated with Al Qaeda, he was regarded as an "unlawful combatant" and was denied entitlements to protection (Liss 2012, 115).

A complication in the Omar Khadr case is that he was not wearing a uniform. This left him open to the charge that he failed to meet an obligation of combatant status: to effectively distinguish himself from the general population (Liss 2012, 116, 121). Recall that legally sanctioned statuses are often associated with maximally public and minimally ambiguous emblems, such as badges and uniforms. There was much disagreement as to whether Khadr had adequately distinguished himself, as Liss (2012) details. A further issue was the question of Khadr's entitlements or obligations as an incumbent of "minor" or "child" status, given that he was aged fifteen when he was captured. Some legal systems specify that a minor is under fifteen, others under eighteen (Liss 2012, 146).

As the Omar Khadr case starkly illustrates, status can be highly ambiguous and contentious and yet highly consequential. Across cultures and societies, we find similar situations. In many societies, a person may be ascribed the status of "witch," "sorcerer," "possessed," or similar—often associated with misfortune, deviance, or mental illness—with sometimes dire consequences such as being ostracized or even killed. In medical contexts, giving someone a diagnosis is giving them a status, with consequences that may include stigma and even institutionalization (Goffman 1963b). The *Diagnostic and Statistical Manual of Mental Disorders*, a formal guide for mental illness diagnosis published by the American Psychiatric Association, is now in its fifth edition. With successive editions, the concepts that underlie various status labels—for example, *addict*—have changed, such that somebody who may be diagnosed as an addict today would not have been so diagnosed thirty years ago. In psychological practice, the concept of *addiction* "has undergone a substantial semantic enlargement in the last few decades" (Haslam 2016, 9). Even those statuses that are regarded as medical conditions, and thus in principle matters of brute reality, are changing at a rapid rate. This underlies the fact that statuses are always, to a large degree, socially constructed.

2.3 Status and Intersubjectivity

Status is integral to human intersubjectivity. We use concepts of status to regulate our own conduct in interactions. Status also provides the tools needed for the *description* of such conduct, both by professional analysts and by participants. And statuses are implicated in understanding any piece of conduct. For instance, you're likely to understand the utterance "He can walk now" differently if it is said by an orthopedic surgeon about her patient or by the same person about her one-year-old son. Many other aspects of context feed into the interpretation of an utterance, but the status of the speaker (and other participants, and people talked about) is crucial.

Statuses can interact with the maxims that people utilize in interpreting ordinary talk and other forms of conduct. For example, a hearer's maxim asserts that "If a category-bound activity is asserted to have been done by a member of some category . . . then hear it that way" (Sacks 1972). Thus, when we read in a book "The baby cried. The mommy picked it up," we should, according to the maxim, understand that the person who picked up the baby was, in fact, the baby's mommy, and that this person was behaving in accordance with that status and not any of the other statuses she may have. So, supposing that the mommy is also a carpenter, to say "The baby cried. The carpenter picked it up" would not be a *false* description of the scene, as it would still describe the participants with applicable reference terms, but it would fail to cohere, as the chosen status term would not thematize the relevant rights and duties that regulate our interpretation of the actor's conduct in relation to the infant.

How is it that "a person comes to grasp and formulate a scene, specifically with respect to the characterization of the actors or participants in it" (Sacks 1972)? Status is key. If we know what behavior is being produced, we can use that understanding to infer a status of the actor. Another possibility is the reverse. If we know the status of the individual, we can use that to understand what action is being done, how, and why. Schegloff (2007b, 473) considers an example from data that Chuck Goodwin collected on an oceanographic research vessel:

> Someone appears on deck with a complicated piece of equipment and says, "Where next?" In the discussion of how to characterize the action this turn was doing—"request for instructions" or "offer of further help"—the issue was recurrently made to turn on who the speaker and addressee respectively were, in hierarchical structure terms. If we could stipulate to the identity of the parties, we could get a solution to the characterization of the action.

The forms of mutual understanding characteristic of human intersubjectivity both presuppose and entail statuses of various kinds. Our account highlights three additional concerns. First, interaction is organized by statuses that extend beyond the immediate boundaries of the interaction itself. Second, many, if not all, occasions implicate multiple, simultaneously relevant statuses. Third, these statuses are "human kinds" in the sense that they are inherently reflexive, being both inhabitable and subject to various kinds of description and redescription.

Take an example from an Indo-Guyanese village in which conduct is pervasively organized through relations of kinship (discussed in Sidnell 2003). There is, for instance, a near universally recognized rule that demands that, *big bodii kyaan taak tu lil bodii waif*, "Big brother can't talk to little brother's wife." This rule points to the way conduct is defined as normatively appropriate for different affinal relationships. For while an older brother is, by

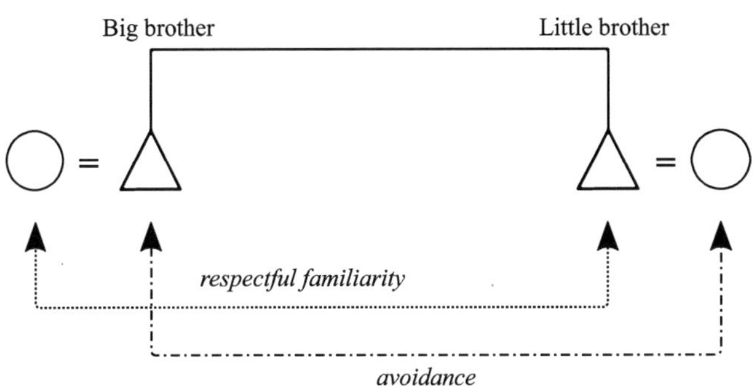

Figure 2.1
Older brother, younger brother, older brother's wife, and younger brother's wife in an Indo-Guyanese village.
After Sidnell (2003).

the rule's injunction, to avoid interaction with his younger brother's wife, a younger man is expected to maintain a relationship of respectful familiarity with his older brother's wife, whom he calls *bhauji*. The term *bhauji* implies a relationship characterized both by positive affect and deference.

Not surprisingly, in a village of fewer than 700 people, older brothers do in fact talk to younger brothers' wives, as a matter of practical necessity. To manage this, an older brother may in fact talk to a younger brother's wife if he does so in a way that indicates that he is not talking to her *in her capacity as his younger brother's wife*. This was the case for two brothers and their wives who jointly owned and managed a small shop. The elder brother in this situation frequently had occasion to talk to his younger brother's wife, but when he did so he restricted his remarks to matters directly pertaining to the immediate operation of the shop. This served to foreground their relative statuses as co-operators of a shop, while backgrounding their status as affines. Similarly, when visiting their younger male siblings, men will often talk to their younger brother's spouse in order to issue directives, such as, "Bring ice now, gyal!", "Come see this thing, pick it up!", and so on. Such directives are not treated as violations of the rule against "talking." Rather, these actions can be seen as instances of normative conduct insofar as the talk in each case is designed to display an orientation to the rule. Although an older brother does indeed speak to his younger brother's wife, the talk is designed so as to show it has been shaped by a knowledge of the rule: its sound is clipped and minimal, and its content is restricted to practical matters that can be seen as necessary in the moment. It is not casual or idle talk, nor talk for talk's sake.

Importantly, though, the rule is sometimes directly and recognizably violated. Villagers are generally organized into patrilocal groups and, within these kin groups, elder brothers hold an authoritative position. If a younger brother's wife is seen to behave inappropriately, in ways that bring shame on the patrilocal group, an elder brother may be called in to talk to her. Similarly, if a woman's husband is behaving in ways that are thought of as inviting censure within the patrilocal group, for example if he is drinking too much, the wife may "give complaint" to his older brother. On such occasions, the visible sense of urgency and importance is conveyed by the fact that such courses of action involve violating the rule as it is stated.

Examples like these clearly illustrate the three points sketched above. First, occasions of interaction are organized, in part through the pervasive application of the rule, by statuses—such as older brother and younger brother's wife—which extend beyond the current encounter. Second, other statuses, such as host and guest, speaker and addressee, and news deliverer and news recipient, are simultaneously relevant both to the production and organization of conduct and its interpretation. Third, these statuses may not only be inhabited and implicitly oriented to but also made the object of discussion and debate. For instance: Should these statuses apply? Are these people incumbents of these categories? Which rules should apply? When and how? Which should be prioritized when they appear to contradict one another?

2.4 Conclusion

We have gone to some length to elucidate the concept of status, the central measure of personhood. Statuses define and organize the normative frameworks that are intersubjectively assumed in our societies. They enter into our interpretations of others' behavior; the planning and execution of our own behavior; and the discourses through which we socially coordinate around people's actions and relationships, maintain shared values and stances, create and destroy reputations, and justify or explain social actions. For these reasons, statuses are fundamental to the human intersubjectivity that social life requires. While any discussion of status may imply a static set of values and relationships, we must emphasize that statuses only exist by virtue of our dynamic orientation to them in the enchronic frame of social interaction. We turn now to a fundamental prerequisite that makes such dynamic orientation possible: *participation*.

3 Participation and Reciprocal Involvement

Any form of engagement presupposes statuses in the sense that we have just described. For instance, at a dinner party, you may have the status of host or guest (or caterer, host's parent who happens to drop by, etc.). And, at various points during the course of the party, you may have the status of speaker or addressee, of storyteller or recipient, and so on. Statuses may be nested and overlapping, and each one may become relevant at different moments and in particular sub-activities they are associated with.[1] But one status is omnirelevant to any form of social engagement: that of *participant*. In other words, for any form of social engagement, we can ask who (or what) has status as a participant within it. Participant is the foundational status for social engagement, and more specifically, for social interaction.[2]

Erving Goffman proposed that whenever people talk together they establish and sustain a world of shared attention and involvement, what he once called a "communion of reciprocally sustained involvement" (Goffman 1957, 49). Participation in an activity is never reducible to mere copresence, contact, or any other set of brute facts. Rather, participation is achieved in relation to norms and is organized moment-by-moment, incarnate, in and through whatever activity people are engaged in. For talk specifically there are practices of speaking, listening, gesturing, posturing, and so on by which people constitute themselves and copresent others as participants in social interaction and participants of a particular kind. A key point for present purposes is that these practices (and the norms by which they are governed) are not simply or solely "regulatory" mechanisms. On the contrary, these practices of participation are foundational to all forms of intersubjectivity including the most developed form, associated with human talk in interaction. We argue that such importance is obscured by any account that conceptualizes the relevant phenomena in terms of mere contact.

In this chapter we take Goffman's argument further. It's not just that the participant is managed in and through practices of interaction. It's also that participation is mediated by larger cultural ideologies of what it means to engage with another person and, ultimately, what a person is in the first place. At the same time, participation in *linguistic* interaction is shaped by the lexical and grammatical means for referring to, and so identifying, a speaker, an addressee, and other nonaddressed yet copresent persons. This means that for language-using humans even the apparently foundational practices of participation have a special, semiotically mediated character and are subject to processes of reflexive reanalysis and revision.

3.1 Phatic Communion and Contact

In "The Problem of Meaning in Primitive Languages" (1923), Bronislaw Malinowski developed a theory of language in which the meaning of an utterance derives from its purpose. Malinowski grounds his theory in a discussion of talk in task activities (such as fishing) but notes that there are types of speech that do not fit this mold. There is talk for talk's sake or, as he put it, "the exchange of words almost as an end in itself" (1923, 312), including "a flow of language, purposeless expressions of preference or aversion, accounts of irrelevant happenings, [and] comments on what is perfectly obvious" (314). This type of speech he termed *phatic communion* (see Zuckerman 2016).

Roman Jakobson later incorporated Malinowski's ideas into his model of speech functions (Jakobson 1960). His discussion of the phatic dimension is brief (Jakobson 1985, 115):

> There are messages primarily serving to establish, to prolong, or to discontinue communication, to check whether the channel works ("Hello, do you hear me?"), to attract the attention of the interlocutor or to confirm his continued attention ("Are you listening?" or in Shakespearean diction, "Lend me your ears!"—and on the other end of the wire "Um-hum!"). This set for CONTACT, or in B. Malinowski's terms PHATIC function (1923) may be displayed by a profuse exchange of ritualized formulas, by entire dialogues with the mere purport of prolonging communication. Dorothy Parker caught eloquent examples: "'Well!' the young man said. 'Well!' she said. 'Well, here we are,' he said. 'Here we are.' she said, 'Aren't we?' 'I should say we were,' he said, 'Eeyop! Here we are.' 'Well!' she said. 'Well!' he said, 'well.'" The endeavor to start and sustain communication is typical of talking birds; thus the phatic function of language is the only one they share with human beings when conversing with them. It is also the first verbal function acquired

> by infants; they are prone to communication before being able to send or receive informative communication.

Jakobson thus used the term *phatic* to highlight the importance of communicative contact, whereas for Malinowski the focus is on the social relationships involved and the purposes to which language is put. Jakobson introduced another crucial modification in suggesting that a given instance of speaking is never categorically or singularly phatic, referential, metalingual, conative, or whatever else. See figure 3.1.

Rather, all of Jakobson's six functions are present together in any bit of talk, though their relative importance varies (Jakobson 1960, 22):

> Although we distinguish six basic aspects of language, we could, however, hardly find verbal messages that would fulfill only one function. The diversity lies not in a monopoly of some one of these several functions but in their different hierarchical order. The verbal structure of a message depends primarily on the predominant function. But even though a set (Einstellung) toward the referent . . . , briefly the so-called REFERENTIAL, "denotative," "cognitive" function—is the leading task of numerous messages, the accessory participation of the other functions in such messages must be taken into account by the observant linguist.

Jakobson's key move then was to rethink the matter in terms of concurrent functions rather than categorical utterance or activity types. But because he continued to think in terms of "predominant functions" that shape "verbal structure," the consequences of this remain muted. Consider some of Jakobson's examples from the passage on the phatic function.

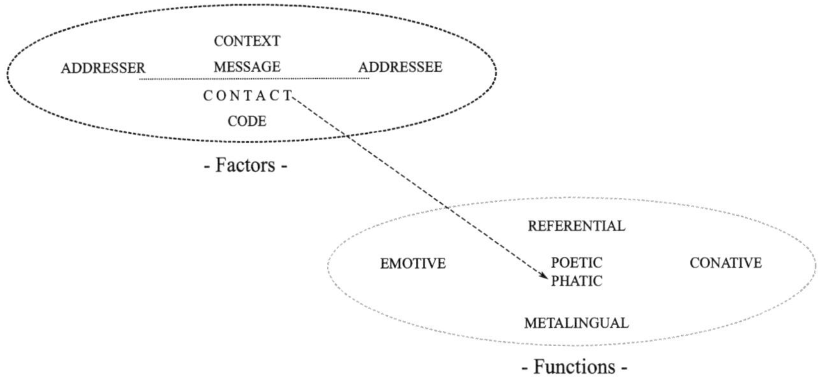

Figure 3.1
Jakobsonian mapping from factors of the speech context to functions of language. After Jakobson (1960).

Jakobson suggests that there are utterances designed "to attract the attention of the interlocutor or to confirm his continued attention," and proposes that "Are you listening?" might be an example. But as studies by Goodwin (1980, 1981) show, such functions are more typically carried out using same-turn self-repair and especially turn-restarts. In the examples Goodwin collected, a speaker perceives that the intended listener is not attending, at which moment the speaker cuts off their speech mid-turn. This typically secures the listener's attention so that the speaker may then restart once the channel is clearly open. By contrast with this off-record method of securing attention, if a speaker says, in so many words, "Are you listening?", this will likely be interpreted not as merely seeking confirmation of the interlocutor's continued attention but as something stronger in addition, such as reprimanding the listener for a suspected lapse. Jakobson did not distinguish between what is embodied in talk ostensibly directed toward quite other ends (e.g., various forms of self-repair) from what is announced and thereby thematized (e.g., an explicit question such as "Are you listening?," which is bound to invite additional inferences).[3]

Despite their insights, Malinowski and Jakobson conceptualized participation in overly simple terms, as a matter of establishing contact or opening a channel. To be sure, this was a significant improvement on the Saussurean notion of a speech circuit, famously diagrammed in the *Course in General Linguistics* (see figure 3.2), insofar as it made the means by which contact is established a possible topic of investigation. Yet it still put forward a massively oversimplified view of the phenomenon.

The reductionism inherent in notions of contact and channel has two aspects. First, this framing results in a binary conceptualization. Contact is either established or it is not and, once established, it is assumed to remain

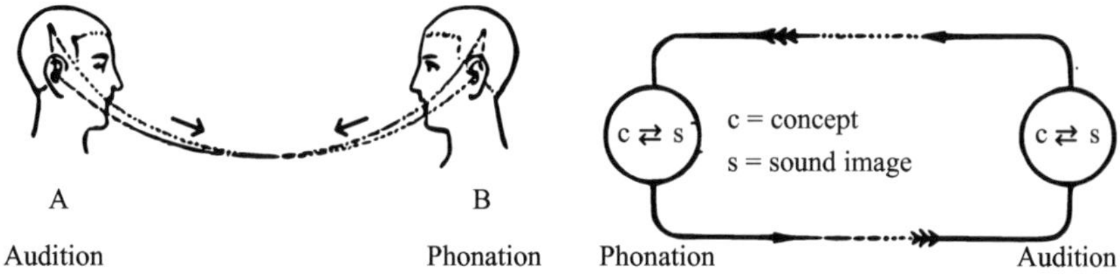

Figure 3.2
Two representations of the Saussurean speech circuit.
After de Saussure (1916).

in the "on" position until there is reason to think otherwise. Similarly, a channel is either open or closed. Second, with the notion of contact, Jakobson failed to appreciate that this is not merely a matter of brute fact (by which contact makes sensory perception possible) but of normative accountability. As Kockelman (2010, 408) puts it,

> Whereas Jakobson understood channels to turn on physical conduits and psychological connections, they should also be understood as turning on social conventions. As an example, we may turn to what is perhaps the most emblematic of semiotic events: joint attention, or looking where another points. Such an event has three key components: a sign (your gesture that directs my attention), an object (whatever you are pointing to: say, that pen over there), and an interpretant (my change in attention).

Kockelman clarifies that here the channel is precisely what allows the "signer and interpreter . . . to intersubjectively relate . . . within a relatively isolated event." And he goes on to note that this channel permits multiple framings (2010, 408):

> In part, this channel turns on a physical contact (e.g., a transparent medium in an illuminated enclosure, with open lines of sight, etc.). In part, it turns on a psychological connection (e.g., I treat your movement as an intentionally communicative gesture, I desire to know what you desire to make known to me, etc.). And, in part, it turns on a social convention (e.g., who is normatively permitted to direct whose attention, in what kinds of contexts, to what kinds of objects).

This last framing of channel—as a piece of normative infrastructure—is key to our argument in this book. It is also the aspect of contact that Goffman pointed to in his studies of social interaction.

3.2 Involvement

In his earliest writings, Goffman developed a radically alternative approach to these matters that begins not with language or speech per se but with social interaction. Goffman conceptualized participation not in terms of contact but attention and, by extension, involvement. For Goffman, then, each episode of interaction, each encounter, constitutes a locally coordinated system in which there is "a single, albeit moving, focus of visual and cognitive attention" (Goffman 1964, 135). The issue for human interactants, then, is not one of making contact or opening a channel, but rather of establishing and sustaining attention to, and mutual involvement in, jointly managed activity. Moreover, we speak in social situations, environments of "mutual monitoring possibilities, anywhere within which an individual will find himself accessible to the naked senses of all the others who are 'present,' and similarly find them accessible to him" (Goffman 1964, 135). Within the general category of social situations, Goffman distinguishes focused from unfocused interactions. The former, which include "card games, ball-room couplings, surgical teams in operation and fist fights," are characterized by a "mutual openness to all manner of communication" and involve "an ecological huddle wherein participants orient to one another and away from those who are present in the situation but not officially in the encounter" (Goffman 1964, 135).

For Goffman, then, speaking takes place within focused interactions or *encounters* and these environments for talk are constituted through a describable set of practices. Goffman speaks here of "functional gestures," which, as the passage quoted below makes clear, is a cover term for practices that include hand gesturing and body-positioning as well as intonation and prosody:

> Utterances must be presented with an overlay of functional gestures—gestures which prop up states of talk, police them, and keep these little systems of activity going. Sounds are used in this gestural work because sounds, in spoken encounters, happen to be handy; but everything else at hand is systematically used too. Thus many of the properties of talk will have to be seen as alternatives to, or functional equivalents of, extra-linguistic acts, as when, for example, a participant signals his imminent departure from a conversational encounter by changing his posture, or redirecting his perceivable attention, or altering the intonation contour of his last statement. (Goffman 1964, 136)

So, rather than the abstract Saussurean speech circuit, Goffman proposes a dynamic system of interaction, which is maintained moment-by-moment by "all manner of communication." From this perspective, de Saussure's speech circuit and Jakobson's notions of contact and channel are reductive abstractions. They gloss over and obscure the practices of interaction that are prerequisite to any speech in context. Moreover, such notions and the conceptual framings of interaction that they implicate fail to take into account the ways in which utterance and context are interwoven and coconstituting.

Such insights led Goffman to see interaction as always involving a delicate balance of competing demands. To maintain the fiction and the feel of ease, each participant must do their part, attending to the right things at the right moments, conveying the right degree of involvement. In "Alienation from Interaction," Goffman describes this balancing act:

> I want to consider the ways in which the individual can become alienated from a conversational encounter, the uneasiness that arises with this, and the consequence of this alienation and uneasiness upon the interaction. Since alienation can occur in regard to any imaginable talk, we may be able to learn from it something about the generic properties of spoken interaction. (1957, 47)

Goffman argues that in order for interaction to come off smoothly, participants must satisfy a variety of "involvement obligations." Part of this requires that a participant pay attention in just the right measure at just the right moments to just the right objects of interest. The problem is that one cannot fulfill such obligations self-consciously, for as soon as one does, attention is redirected from the course of action underway toward the problem of being spontaneously involved in it.

Goffman describes four ways in which a participant may become alienated from interaction:

1. ***External preoccupation:*** "The individual may neglect the prescribed focus of attention and give his main concern to something that is unconnected with what is being talked about at the time and even unconnected with the other people present, at least in their capacity as fellow-participants" (1957, 49).
2. ***Self-consciousness:*** "At the cost of his involvement in the prescribed focus of attention, the individual may focus his attention more than he ought upon himself—himself as someone who is faring well or badly, as someone calling forth a desirable or undesirable response from others" (1957, 50).
3. ***Interactional consciousness:*** "A participant in talk may become consciously concerned to an improper degree with the way in which the interaction, qua interaction, is proceeding, instead of becoming spontaneously involved in the official topic of conversation" (1957, 50).
4. ***Other-consciousness:*** "During interaction, the individual may become distracted by another participant as an object of attention—exactly as in the case of self-consciousness he can become distracted by concern over himself" (1957, 51).

In this early work, Goffman described interaction from the point of view of the individual and focused on the causes and consequences of alienation for individual participants. But his insights also apply to interaction framed dyadically. With the idea of focused encounter as a collaborative achievement we can consider the practices by which participation is constituted in face-to-face interaction. This is particularly clear in cases where there is a question as to whether some individual is or is not a participant.

Consider the following fragment of interaction from a backyard barbecue in which three men (Curt, Gary, and Mike), one child, and one woman (Carney) are seated at a table. The men have been telling dirty jokes. In figure 3.3, Curt, Gary, and Mike are arranged so as to form a focused encounter—an environment of mutual monitoring possibilities—while Carney sits with her back to the table, facing in the other direction. This organization provides for a kind of equivocal participation insofar as Carney is present and can hear what the others are saying without necessarily counting as a ratified recipient of the talk at the table.

The participants orient to this equivocality. At one point, Curt, who is Carney's cousin, is telling a joke to Gary, which he presents as a true story, about the time that he and Mike were captured by the enemy in Vietnam. Gary, the primary addressed recipient of the joke, expresses doubt, saying,

Figure 3.3
Equivocal participation: Carney is at the table but has her back turned.
After Goodwin (1986); Sidnell (2011).

"Mike an' I?" and then "Oh horseshit." In response to these displays of skepticism, Curt attempts to enlist Mike with "Didn't we. Remember Mike?" At this point, Carney turns to her cousin Curt and says, "You forget I'm here." By saying this, Carney is suggesting that she can disconfirm what Curt is saying. Indeed, Curt, who has persevered in the face of multiple challenges by Gary, here appears to acquiesce to Carney. Not only does he suspend the telling in progress, saying "Alright," he joins with Carney in a bout of laughter. In this way, Curt displays that he has been "caught out" by Carney and does not in fact proceed with the telling until Carney instructs him with, "Go on." Notice, then, that in formulating this in the way she does, Carney characterizes her own participation as something that has been treated by others as equivocal, ambiguous, or peripheral—as someone who, though present, can be, and has been to this point, disattended and, in fact, forgotten. A closer study of this episode (Sidnell 2011) argues that this equivocality is carefully managed by the participants, and especially by Carney, in such a way as to avert potential recipient-design problems implicated in the telling of dirty jokes (again illustrating the normative nature of contact).

This example shows that a participant's status is always relative to some frame of activity. So, for the activity that can be glossable as "having drinks at a backyard barbecue," Carney is a ratified participant, whereas for the activity glossable as "the telling of dirty jokes," Carney's participation is equivocal. Participation is calibrated to such nested activities, and people in interaction display an orientation to this order of organization.

To take another example from the same occasion, Curt, Mike, and Gary are talking about local car races and the drivers who compete in them. These men apparently do not share this interest with the women who are also at the barbecue. The women do not actively participate in the discussions about the races and the men make no effort to include them. During one stretch of talk about a driver named Fisher, Phyllis, who is Mike's wife, says, apparently to no one in particular, "Need more ice," picking up her glass and an empty bottle of Coke as she stands up and walks away from the table. Why does she make this unilateral assertion, with no obvious connection to the interaction taking place? Goodwin (1986) analyzes the utterance as an account for Phyllis's action, a way to block any inference that, by departing from the table, she is reacting to what has been said or perhaps to the fact that she has not been included in the conversation. Our point is the following. If her utterance seems not obviously connected to the interaction at hand, it is because we are framing her in relation to the men's conversation about the race car driver named Fisher. If instead we frame her action in relation to the activity of "having drinks at a backyard barbecue," then we can readily understand her as a participant, inhabiting a status with certain rights and duties, and in that light her utterance upon leaving the table can be understood as an account, tied to her status within the second activity-frame.

3.3 Cultural Conceptualization and Social Interaction

Goffman begins "Alienation from Interaction" with the assertion that "when the individual in our Anglo-American society engages in a conversational encounter with others he may become spontaneously involved in it" (1957, 47). This suggests that his claim pertains to a particular (ethnically defined) social group. But he does not dwell further on the matter and thus does not address the larger issues that this suggestion necessarily raises. For instance, could there be substantial differences across social groups in the organization of interaction and mutual attention? If so, what could account for those differences? Would they be explained as the effects of cultural ideologies, social organization, or what?

Unfortunately, Goffman has no theory of cultural conceptualization. When he uses the word *culture* it is almost always in reference to "a culture" or "cultures" in the plural, something akin to a social group among whom conventions are shared. He does not think of culture in terms of widely circulating, if socially distributed, metasigns. For him, metasemiotic activity is essentially limited to immediate accountings and somewhat more distal reports (of news items, for instance, of which he was an avid collector). His analysis is exceedingly rich and nuanced and his eye for interactional strategy unrivalled, but the pervasive need to qualify his findings (*à la* "our Anglo-American society") points to a significant theoretical and analytical limitation. It is theoretically limited because such a caveat at the outset blocks important comparative questions, as well as any causal account. And it is analytically limited because it allows Goffman to avoid addressing the relation between what people do in interaction and the culturally specific ways in which they talk and think about what they do.

It is instructive to compare Goffman's approach with a quite different, though contemporaneous, one. While Goffman was studying alienation from interaction, Clifford Geertz was conducting the research that would be published in 1960 as *The Religion of Java*. This, Geertz's stunningly rich and nuanced first ethnography, includes a chapter on interaction in Java and especially among the elite bureaucratic *prijaji* class.[4]

According to Geertz, Javanese culture, or at least Javanese conduct, is organized around two sets of opposed concepts. On the one hand, *alus* "smooth, refined, elegant" is opposed to *kasar* "rough." On the other, *batin* "inner" is opposed to *lair* "outer" (see also Errington 1984). Geertz writes:

> *Alus* means pure, refined, polished, polite, exquisite, ethereal, subtle, civilized, smooth. A man who speaks flawless high-Javanese is *alus*, as is the high-Javanese itself. A piece of cloth with intricate, subtle designs painted onto it is *alus*. An exquisitely played piece of music or a beautifully controlled dance step is *alus*. So is a smooth stone, a dog with his hair petted down, a far-fetched joke, or a clever poetic conceit. God is, of course, *alus* (as are all invisible spirits), and so is the mystical experience of Him. One's own soul and character are *alus* insofar as one emotionally comprehends the ultimate structure of existence; and one's behavior and actions are *alus* insofar as they are regulated by the delicate intricacies of the complex court-derived etiquette. . . . *Kasar* is merely the opposite: impolite, rough, uncivilized; a badly played piece of music, a stupid joke, a cheap piece of cloth. . . . Between these two poles the prijaji arranges everyone from peasant to king. (Geertz 1960, 232)

Geertz explains that this opposition needs to be understood in relation to that between inner and outer:

> *Batin* means "the inner realm of human experience," and *lair* "the outer realm of human behavior." . . . *Batin* refers not to a separate seat of encapsulated spirituality detachable from the body but to the emotional life of the individual taken generally—what we call "the inner life," or "the subjective"; it consists of the fuzzy, shifting shapes of private feeling perceived directly in all their phenomenological immediacy. . . . *Lair*, on the other hand, refers to . . . external actions, motions, postures, and speech of the individual. The ordering of the outward life leaves one free to turn to the ordering of the inward. The cultivated man needs to give form both to the naturally jagged physical gestures which make up his external behavior and to the fluctuating states of feeling which comprise his inner experience. A truly *alus* man is polite all the way through. (Geertz 1960, 232)

The *prijaji* aims, then, to cultivate *alusness* both in his inner, emotional life and in his outer conduct.[5] According to Geertz, the *prijaji* sees this as a single coordinated task pursued through the means of meditation on the one hand and etiquette on the other. The result, if successful, is the cultivation of "emotional equanimity, a certain flatness of affect." To achieve this "prized psychological state" (Geertz 1957, 431), the *prijaji* seeks to limit exposure to any shock from the external world and in this he has "two lines of defense . . . the formalization of social behavior so that it is easily predictable, and the flattening of affect so that if the unexpected does occur it will be less likely to lead to an inward disturbance." (Geertz 1960, 241)

Alus and *kasar*, in Geertz's description, conform closely to what Bernard Williams ([1985] 2011, 155 and passim) calls "thick ethical concepts": highly salient terms that combine a descriptive and an evaluative component. Much of the philosophy of ethics has, according to Williams, adopted a largely reductive approach to such concepts, attempting to isolate the "is" from the "ought." Importantly for our purposes, Williams argues that such concepts often provide "someone with a reason for action," and that though their "application is guided by the world" in the sense that they can be used "rightly or wrongly," they are at the same time "action-guiding." (Williams [1985] 2011, 155–156) In other words, these are the concepts in terms of which people think and talk about what they do, and that they use in accounting for their own and other people's behavior.

In the Javanese case, *alus* and *kasar* have a pervasive relevance in the flow of interaction. Because the goal is to cultivate *alusness* both in the inner and the outer realm, participants in interaction attempt not only to smooth out their own behavior (so as to display their own *alusness*) but also to avoid doing anything that would disturb others. Geertz documents several strategies directed toward this effort (see, for instance, his discussion of *andap-asor* and *étok~étok*) but most famous is the elaborate system of linguistic honorification conveyed by different speech styles.

The structural features of these styles, known as *ngoko*, *madya*, and *krama* (Geertz 1960) or *ngoko* and low and high *basa* (Errington 1998) have been well-described elsewhere. Our concern here is how practices of etiquette are talked about and rationalized by those who use them. To understand this, we have to begin with self-directed disciplines of meditation by which a *prijaji* hopes to calm "inward feelings" and, ultimately, to "build a wall around them" such that one is "able both to conceal them from others and to protect them from outside disturbance." (Geertz 1960, 241) In the interpersonal domain, the wall is built with etiquette.

> The wall, then, is a wall of etiquette. Etiquette provides the *alus* prijaji with a set of rigidly formal ways of doing things which conceals his real feelings from others. In addition, it so regularizes behavior, his own and that of others, as to make it unlikely to provide unpleasant surprises. . . . Politeness is something one directs toward others; one surrounds the other with a wall of behavioral (*lair*) formality which protects the stability of his inner life (*batin*). . . . Etiquette is a wall built around one's inner feelings, but it is, paradoxically, always a wall someone else builds, at least in part. (Geertz 1960, 241–255)

Geertz goes on to illustrate the construction of such a wall through linguistic etiquette with a series of elegant diagrams (see figure 3.4, below) in which the solid center represents the inner *batin* and the concentric circles the stylistic levels (solid lines) and optional honorifics (dotted lines) used in speech addressed to another person. Lower-ranking people are required to speak "up" to their superiors who in turn speak "down" in response. The A diagram in figure 3.4 represents a "high official" such as a district officer and an "ordinary educated urbanite." In a case such as this, "the District Officer will speak *ngoko biasa*, the ordinary man, *krama inggil*." (Geertz 1960, 256) The B diagram represents a peasant speaking to a higher-status person. The peasant will use *krama madya* since he lacks competence in the higher registers of *krama biasa* or *krama inggil*. Finally, the C diagram represents one *prijaji* speaking to another. The increasing formality and formalization of speech as one moves from the relatively free and easy styles of *ngoko* to the highly structured styles of *basa* has the effect of regulating both *what* is said and *how* it is said. According to Geertz, the more elevated the speech style the less propositional information is conveyed. In other words, semiotic form serves to regulate, and, to some extent at least, constrain the expression of content (see also Bloch 1975). In relation to the argument of this book, one might see elevated Javanese speech styles as something like uniforms in that

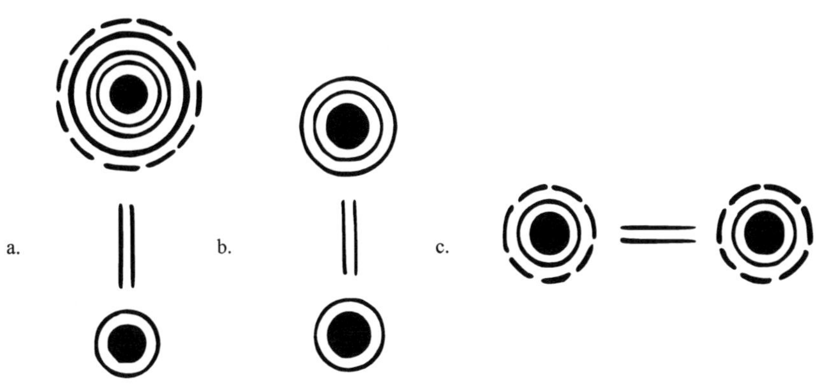

Figure 3.4
Javanese speech patterns. (a.) High official speaking to an ordinary educated urbanite, (b.) peasant speaking to a higher-status person, (c.) one *prijaji* speaking to another using *ngoko sae*. The central dots represent individuals and the rings around them represent the linguistic embellishments that must be added when speaking with/about that person.
After Geertz (1960).

they both emblematize status and, at the same time, narrow the range of what can be done, thereby making conduct all the more predictable.

We hope that this brief discussion of Javanese interaction brings out a clear contrast with Goffman's picture of "Anglo-American society." The point is not that the underlying framework of interaction differs. Rather, the differences lie in the "culture of language use" (Silverstein 1979, 1985), that is, the various ways in which people talk about and rationalize their own and others' linguistic behavior. Goffman's Anglo-American attaches great importance to spontaneous involvement, while in Java restraint and formality are the goal. As Geertz notes (see also Seigel 1986), the cultivation of *alusness* through etiquette figures in the very definition of what it means to be a (Javanese) person.[6]

> According to the Javanese, the difference between men and animals is that the former "know order." By "order," the prijaji means formality of bearing, restraint of expression, and bodily self-discipline—a constant awareness of himself as being an object of perception for others and therefore obligated to present a pleasing, *alus* picture. *Spontaneity or naturalness of gesture or speech* is fitting only for those "not yet Javanese"—i.e., the mad, the simple-minded, and children. (Geertz 1960, 247, emphasis added)

In sum, participation in Javanese interaction, as Geertz describes it, would seem to involve, as a prerequisite, exactly the kind of "interactional consciousness" that Goffman suggests inevitably results in alienation.

3.4 Avoidance

Practices of interpersonal avoidance carry this "antiphatic" theme further (Fleming in press). Avoidance techniques range widely. There are subtle restrictions on patterns of address observed in many English-speaking communities (e.g., *Kim* versus *Mrs. Smith*; Brown and Ford 1961; see also Fleming and Slotta 2018). In Aboriginal Australia, there are much more elaborate and extensive prohibitions that constrain and curtail virtually all aspects of interaction between certain categories of person (Haviland 1979; Merlan 1997). Similar practices are observed in small-scale societies worldwide. Consider the kinds of avoidance practiced and indeed normatively required among the Yomut Turkmen (Irons 1975) or the Buryat Mongols (Humphrey 1993). Among the latter, a newly married woman, living in her husband's camp, "was subject to an extraordinary number of regulations in dress, comportment, orientation and the like . . . One set of rules was concerned with maintaining a calm, almost blank demeanor. She was not allowed to raise her head, to laugh, to cry, to sing without being asked, to grind her teeth, or make any noise when eating." (Humphrey 1993, 84) Such taboos severely restrict the newly married woman's capacity as a signifying agent and thus her participation in social interaction. At the same time, a range of prohibitions prevented any expression of her sexuality, especially when in the presence of her father-in-law. Thus, she was required to always "wear a hat in his presence, do up all her buttons, pull down her sleeves over her wrists," and the father-in-law in turn was expected to ignore her. Any instructions he might issue were transmitted as commands through the mother-in-law (Humphrey 1993, 84).

Such practices of avoidance have the effect of limiting, curtailing, or diminishing an individual's semiotic agency (Kockelman 2017). Among the Buryat, not only was the wife's conduct severely curtailed, but others were expected to, inasmuch as possible, ignore her, thereby disattending those signs she nevertheless produced. Among the Yomut Turkmen the prohibitions go further, limiting the new bride's capacity not only to produce signs but also to perceive them. Irons (1975, 105) writes:

> The norms relating to a woman's behavior toward her affines are extremely elaborate. . . . During her earliest contacts with her father-in-law she throws a cloth over her head completely covering her face. . . . While veiled in this way a girl cannot see and needs someone to guide her. . . . In the early period of residence in her father-in-law's household, she must veil herself in this manner from all members of the household senior to, or equal with, her husband including wives of older brothers, and from her husband as well when any of her husband's seniors are present.

As Fleming (in press) notes, this and other veiling techniques have an "antiphatic" function, imposing obstacles to mutual access. In many of the more familiar cases of dress such as the burqa or chadaree, the veil obstructs others' access to the one who wears it. In the case of the Yomut Turkmen access is restricted in both directions. Fleming (in press, 2) remarks that, among the Turkmen, the "most extreme veiling not only occludes the head completely, it blocks off the visual channel, rendering the signaling subject literally dependent upon a member of the conjugal household to move her through domestic space."

After some time, the wife adopts a diminished head covering known as the *bürünmek*, which "consists of covering the face below the eyes with a cloth, and then pulling one's head cloth down over one's forehead in such a way that only a slit is left open for the eyes" (Irons 1975, 106). After the birth of a first child the wife adopts a third style of veiling, called *yashmaq*, which "entails concealing one's mouth and not speaking to, or eating from the same dish, as the person avoided" (Irons 1975, 105). As Fleming (in press) notes, beginning with marriage and concluding with the birth of a child, "the bride's progressive incorporation into her husband's family is sketched out by a scaled reduction in avoidance practices." Moreover, these veiling styles are "scaled with respect to one another" (4).

> The cloth over the head hides the eyes and the mouth from others and makes seeing impossible. The *bürünmek* liberates the eyes for seeing while carefully protecting them and the mouth from being seen. Finally, with *yashmaq* only the mouth remains covered. This scaling figurates degrees of phatic accessibility. In total covering, the bride is forestalled from co-presencing, in any way, the members of her husband's household. In the *bürünmek* hiding of her eyes, she studiously forestalls eye contact, closing off a crucial means by which interactional co-presence is mutually ratified, itself an important scaffold for face-to-face communication. Lastly, the covered mouth of *yashmaq* represents the asymmetric closure of the speech channel; though she can now be presenced and even interpellated, the bride cannot initiate communication with her parents-in-law or channel communication directly to them, but must indirectly—through non-verbal gesture or whispered use of speech intermediaries—respond to their entreaties. (Fleming in press, 4)

The Javanese and Turkmen cases point to the way participation in interaction is culturally construed and inflected through a variety of semiotic means. What is handled by "walls" of etiquette and techniques of the self in the Javanese case is managed with physically obstructive shields among the Turkmen. And both cases reveal the extent to which Goffman's observations were historically and group specific. The environment of "mutual monitoring possibilities" is not the brute fact of copresence, the baseline for human interaction, but rather a cultivated outcome and a normatively prescribed form of behavior. It is one among many alternatives observed in cross-cultural perspective.

Such practices of avoidance, like the Javanese patterns, do not negate intersubjective engagement so much as they shape it while reflexively characterizing its significance.[7] In discussing the avoidance techniques that sons-in-law and mothers-in-law practice among the Korowai of West Papua New Guinea, Rupert Stasch (2009) writes:

> erasing an affinal counterpart as an object of perceptual contact does not erase that person as an interactant. Rather, *avoidance is a form of intense reciprocal engagement*. A pair's practices of separation brusquely intrude on everything else in a

> social scene. Other activities they undertake are dominated by their vigilance in staying out of each other's sight. (87, emphasis added)

This echoes Humphrey (1993) who, in reporting on the extensive and elaborate avoidance practices that a woman is expected to maintain in the presence of her in-laws, suggests that "although phrased in the negative," taboos "always and necessarily imply some kind of positive action" (Humphrey 1993, 73):

> If one act is forbidden, people do something else. The existence of a taboo means that behavior in that particular field of action is "marked" or significant, and therefore the acts which substitute for taboos, or are performed instead of them, should not be seen merely as "ordinary" behaviour, as opposed to tabooed behaviour. (Humphrey 1993, 73)

Indeed, the presence of a taboo suggests a cultural preoccupation with that domain of conduct that is prohibited or regulated by it (see Foster 1966). Among the Buryat, an in-marrying woman must avoid pronouncing the name (or title, e.g., "horse-herdsman") of any of her senior male affines and so, given that such names are frequently homophonous with ordinary, commonplace words, must make frequent circumlocutions and substitutions that cause her talk to sound avoidance-inflected. This is in addition to the various forms of attentional and physical avoidance she must practice. This has consequences for status and bears in a key way on our discussion of status in chapter 2. The in-marrying wife becomes, by virtue of the taboos that regulate every aspect of her behavior in the presence of others, inescapably tied to this one possible status. It is impossible for her to engage with others on any other footing than as "in-marrying woman."

Merlan (1997, 108) makes this point in reference to the anthropology of Aboriginal Australia, noting that

> full enactment of the mother-in-law avoidance taboo, in which the highly prescribed and institutionalised forms—aversion of gaze, relative taciturnity, special proxemics, and so on—tend to give *one* aspect of the relationship between people overriding determination of their conduct in each other's presence. Thus, a man in his mother-in-law's presence finds it difficult to behave towards her in any way other than as her son-in-law; and further, his conduct towards everyone else on the scene is very strongly shaped by their co-presence, and the social emphasis placed on it.

As Stasch (2009, 87) puts it of the Korowai, when such a pair are "in each other's vicinity, the affinal relation surrounds each of them as an intangible, invisible, and transcendent presence."[8]

The examples so far have focused on ways in which practices of social avoidance restrict or lessen opportunities for participation, often with reference to kin relations. Another logical possibility is that avoidance behavior *redirects* rather than reduces opportunities for interactional participation. An example comes from Kri speakers of the Nam Noi valley in upland central Laos. In this community, a person's opportunities for participation are differentially distributed by means of norms that specify who should sit where in the physical layout of houses (Enfield 2009; Zuckerman and Enfield 2022). Upon marriage, young men find themselves the subject of a specific set of strictures, one of which is that when they are in the house of their parents-in-law—their main domicile in the early years of marriage—they are not permitted to remain in the area of the house usually reserved for men and guests. Rather, as a son-in-law, they must remain in the area reserved mostly for women and children. This doesn't so much reduce their possibilities for participation as it directs them to a new section of people in the social milieu; namely the women and children of their affinal household.

A final kind of culture-related constraint on participation doesn't constrain who you can talk to but rather whether (or how) you may talk at all. In many Aboriginal Australian societies, mourning is associated with silence. People whose loved ones have recently died may avoid speaking at all for a period (Dixon 2002).[9] Most if not all cultures have some proscriptions against participation as a matter of decorum. In Western Apache society, silence (i.e., nonparticipation) is the norm in certain situations, for example when meeting strangers, courting, being with people who are sad, getting cussed out, or when children return home (Basso 1970).[10] These situations have something in common, grounded specifically in the matter of status: "In Western Apache culture, the absence of verbal communication is associated with social situations in which the status of focal participants is ambiguous" (Basso 1970, 96). We would suggest it's not just whether the status is ambiguous—statuses always are, in the sense that we have many at once—but rather that participants want to explicitly orient to that fact. The silence is, after all, communicative. These are situations that, universally, tend to require that people be silent or still. There are even claims that certain cultures place a general value on a reduced degree of participation across the board. For example, it is widely held that Native Americans have "a fierce reluctance to speak except when absolutely necessary" (Basso 1970, 80) and that Finnish speakers value silence in conversation (Lehtonen and Sajavaara 1985).[11]

3.5 Configurations of Speaker and Addressee

Participation in interaction is semiotically mediated not only by such cultural conceptualizations as Javanese *alus* and *kasar* and by semiotic techniques such as veiling among the Yomut Turkmen. Also implicated are the grammatical and lexical resources that are used to refer to, and hence identify, the speaker and the addressee. In English and many other languages this is accomplished by means of pronouns that exhibit a property of *token-reflexivity*: the referent of the pronoun "I" is the speech participant who utters a given token of the word. In Vietnamese and a number of other languages of Southeast Asia, speakers typically avoid the use of token-reflexive pronouns in favor of lexical nouns, most often kin terms (Fleming and Sidnell 2020). One recent study (Sidnell 2022) found that 80 percent of references to speaker and addressee were accomplished by kin terms in a sample of Vietnamese conversations between nonrelated, same-generation peers. Consider the following case in which, while having coffee with their spouses at a café in Hanoi, Huy, a man in his early thirties, asks Thanh, a slightly older man, whether he wants Huy to turn on the TV.

(1) (Vietnamese)

01 Huy: *Em mở cho anh xem*
younger-sibling open give elder-brother watch
"You want me to turn on the TV so (you can) watch?"
02 Thanh: *Đây*
"Yeah."

In this utterly typical and mundane case, the speaker, Huy, refers to himself as *em* (younger sibling) and to the addressee, Thanh, as *anh* (elder brother). Unlike first- and second-person pronouns, which index the participant roles of speaker and hearer, respectively, these terms are not, in terms of grammatical or semantic structure, connected to the participants *as* speech-act participants. Rather, they denote the institutional roles of the family and are here used metaphorically to depict the relationship between the participants according to this model. In order to do so, the identity of speaker and addressee must be already established by other means such as gaze direction, pragmatic inference from what is said, and real-world knowledge about the relative ages of the participants.

When used for interlocutor reference, kin terms like these have two important collateral effects. First, often in a language such as Vietnamese the addressee is not identified by a term that refers to "him" or "her," but rather by implication to the term used in speaker self-reference. For example, when a speaker refers to herself as *em* (younger sibling), she shows that she is addressing the one person present who is older than she, not the several other people who are younger. That is, the choice of speaker-referring expression contributes to *addressivity*. Second, in a language like Vietnamese the distinction between reference and address is not as stark as it is in those languages that rely primarily on pronouns.[12] Although there are particles that, when attached to a common noun like a kin term, unambiguously signal reference to a third party (that is, a person, who, whether present or not, is not an addressee of the utterance), the absence of such a particle does not necessarily imply that the individual referred to is the addressee, and indeed such unmarked forms occur in utterances in which another participant has clearly been selected as the addressee. Thus, the distinction between address and reference—which figures so centrally in the study of kinship terms and, by extension, kinship generally (see Luong 1990 for discussion)—is something of a language-specific artifact. Where unsupported by a robust grammar of person, it is at best a pragmatic implication and not a semantically coded component of the message.

On the ground, in occasions of interaction, this has practical consequences. In Vietnamese interaction there is a way of speaking *about* someone without thereby casting that person as a nonparticipant in the way, for instance, English *he* necessarily does. To see this, we must consider two kinds of cases. The following examples are from the same occasion as (1) above. In example (2), Linh, the spouse of Thanh, is holding up Huy's rather large keychain, selecting individual keys from it and asking Phuong, Huy's spouse, what each of them is for.

(2) (Vietnamese)

01 Linh: *Đây là chìa khoá cử'a nào đấy.*

this is key door which there.

"This is the key for which door?

02 *Ôi, chìa này giống nhà mình nhể*

oh, key this like house self see?

Oh, this key is like my house key, see?"

03 Phuong: *Em đừng hỏi chị*

younger-sibling don't ask elder-sister

"Don't ask me!"

04 *Những-cái-chìa-khoá anh ấy đi đâu,*

the-keys elder-brother this go where,

"Those are his ('this elder brother's') keys to go somewhere.

05 *phòng nào chị không biết hết đâu*

room which elder-sister not know at all

"I have no idea what room they are for."

06 Linh: *Ủa ôi. Phòng nào không biết à*

ah. Room which not know eh

"Don't know what room this is for, eh?" [pointing to a particular key]

07 Phuong: *Không*

"No."

Here, when Linh asks Phuong about a particular key, she responds by disclaiming knowledge on the grounds that they belong to her husband Huy. In doing so she refers to Huy, who is copresent and seated directly across the table from her, using the form *anh ấy*, which combines *anh*, the kin term for elder brother (spouses normatively use sibling terms in referring to one another), with the particle *ấy*, which marks such a kin term as referring to a third person.

Next, in example (3), after Linh presses the issue, suggesting, in jest, that Huy is keeping secrets from his wife, Phuong details what she does know about the keys. In doing so, Phuong tells Linh that Huy gave a complicated explanation for one of the keys that she does not remember. She refers to Huy using the bare kin term *ông* (grandfather), and although she is clearly not speaking *to him*, she does not append the particle *ấy* as she did in line 3 in the last example (*anh ấy*).

(3) (Vietnamese)

01 Phuong: *Biết mỗi chìa-khoá này chìa-khoá xe,*

know each key this key motorbike,

"[I] know this is the key for the motorbike,"

02 *chìa-khoá này chìa-khoá cử'a*

key this key door

"this is the door key."

03 *Còn đây là, ông cứ' giải thích loằng ngoằng lắm*

and this is, grandpa kept explain zig zag

"And this is, he ('grandpa') gave some complicated explanation,"

04 *không nhớ' nổi*

not remember able

"that (I) don't remember."

05 Linh: *Anh, nhữ'ng-phòng nào đây.*

elder-brother, rooms which this.

"Elder brother [Huy], which rooms are these?"

06 *Phòng nhất phòng nhì à*

room first room second eh

"Room of wife, room of mistress, eh?"

By referring to Huy in this way, Phuong attempts, it would seem, to reconfigure the framework of participation, effectively drawing Huy into the interaction, though she is clearly speaking to Linh. And we may note that at the next possible opportunity, Linh realigns the interaction, now addressing her teasing question not to Phuong but to Huy. In the case of Vietnamese, then, we can see the way the grammatical and lexical resources employed in practices of interlocutor reference have consequences for the configuration of speaker and addressee and thus for the organization of participation.

3.6 Conclusion

Participation in intimate, copresent interaction is organized in and through the semiotic resources of language and the body, and in some cases with additional technologies such as the veil. But public participation—participation in publics of various kinds—generally requires a more extensively elaborated infrastructure: literacy, script and mechanical means of print production, smartphones, Wi-Fi networks, servers, and so on (Cody 2011). At one step up from face-to-face communication, forms of amplification can provide for larger audiences but do not readily allow for reciprocal contributions from participants, thus introducing an asymmetry between the speaker and those addressed. An unamplified version—known as the "people's microphone"—involves one person speaking in portions short enough that they can be repeated by those within earshot, and thus heard by those farther away from the original speaker. This is then repeated, if necessary, so as to reach groups at an even greater distance. Interestingly this method of public participation is subject to a wide range of characterizations, from brainwashing to liturgy to "a lesson in the obstinacy required for intentional, durable citizenship."[13]

Such examples, along with the cases of sawing and weaving we started with in chapter 1, push toward a more expansive conception of participation, showing that participation is hardly a simple matter: "the idea of 'being part' has far greater philosophical and anthropological implications than we might have expected at first" (Pina-Cabral 2018).

As a case in point, consider how central to current thinking are ideas about "political participation," "participation in public life," and so on (MacPherson 1962). In our account, these forms of engagement with others involve normative statuses (e.g., citizen, voter)—all of which are social-relational by definition—and moreover give rise to distinct forms of intersubjectivity. Some of these issues have been addressed, in philosophy and later in anthropology, in relation to arguments about the emergence of a public sphere and publics more generally (Habermas 1989; Warner 2002; see also Cody 2011, Gal and Woolard 2001). Habermas describes the emergence of apparently novel forms of copresent interaction—including the salon and the coffeehouse—in which, according to him, for the first time people belonging to a particular group in society (the bourgeoisie) participated, normatively, as reason-bearing agents rather than as actors tied to particular social roles and statuses. These roles and statuses, according to Habermas, were bracketed to allow for the free exchange of ideas (see Fraser 1990). Such forms of public interaction constitute an interesting example within our general account of participation but they do not necessitate a substantial rethinking or modification of the terms of our argument.[14]

But other developments that Habermas pointed to and that Warner and others have discussed in detail suggest some ways in which our account of intersubjectivity—as activity-dependent, emergent in activity, and always tied to and generated by human engagement with others—should be expanded.

Consider the famous story of Martin Luther nailing his "Ninety-Five Theses" to the door of All Saints' Church in Wittenberg on October 31, 1517. Assuming that this is historical fact, we can imagine that by such an act Luther initiated a communicative activity and invited any passersby, and perhaps especially those people associated with the church, to read his arguments against the selling of indulgences. A person might see the notice, read it, and return home and report on it to their family. A member of the family might then discuss it with a friend, and so on. Each link in the chain would be vulnerable to distortions and amplifications. Such an act might set off a wave of discussion and ultimately a protest and an eventual proposal for reformation of the church. Supporting such a process, in this imagined scenario, would be the printing of the theses and their distribution among the literate population. This would no doubt result in further discussion in face-to-face mode of the strengths and weaknesses of Luther's arguments, his entitlement to make them, how they fit with matters of religious duty and piety, and so on.

But consider now the difference that text circulation makes. For if Martin Luther had lived in the 1800s he might have published his theses in a newspaper or other circular, in which case his controversial arguments would have no doubt elicited reply. In those replies the authors would have talked about specific details in Luther's theses, but they would not have, at least not necessarily, addressed themselves specifically to Luther but rather to a reading public.[15]

Such mediated forms of social interaction have different conditions of participation (e.g., literacy), different means of contributing (e.g., letter writing, editorial columns), and result in different forms of intersubjective understanding. But they build on a fundamental architecture of forms of participation such as those we have described here. And participation—in our technical sense of people's access to one another as a precondition for social interaction, and thus for intersubjectivity—shows a degree of sociocultural variability, given that it is causally bound up with interpersonal relationships, which are known to vary dramatically in kind across cultures and societies. By contrast, once participation is secured, we see far less scope for cultural variation in the dynamic mechanics of social interaction itself. This is because social interaction presupposes an infrastructure with certain properties (Enfield and Levinson 2006; Levinson 2006; Enfield 2013; Enfield and Sidnell 2014). That infrastructure—the enchronic infrastructure for human interaction—is the topic of the next chapter.

4 Enchrony and the Semiotic Bottleneck

In this chapter we make good on the promise of chapter 1 to sketch out the architecture of intersubjectivity associated with talk in interaction.

An initial condition for social interaction is that an informational channel must be open. We outlined this in the last chapter under the rubric of *participation*. But once participation is established, and a communicative channel is open, other constraints apply. We can think of these as sieves or filters that limit the possibilities for what can be transmitted across a semiotic channel. We will refer to these sieves and filters—taken together as a constraining set of factors on the realization of sign events in social interaction—as the *semiotic bottleneck*. We note three types of constraint that play a role in constituting this bottleneck.

First, if I am to communicate with you, the signs I use must be *qualisigns* (Peirce 1955; Parmentier 1994; Kockelman 2005). Their physical qualities must be included in the set of qualities that can possibly be perceived by you. For humans, these include speech sounds, facial expressions, and bodily gestures, but do not include stimuli that are outside our range (for example, the ultrasonic sounds that are qualisigns for dogs) or are too subtle to be perceived by us (for example, the pheromones that are qualisigns for ants, or the tiny distant movements of mice in fields that are qualisigns for hovering eagles). The perceptual capacities of any interpreting agent determine the range of possible signs for that agent, delimiting the set of differences that could make a difference in communication (Bateson 1979).

Second, to actually communicate with you, I must use *sinsigns*, that is, materially realized tokens of qualisigns that are actually perceived and reacted to in semiotic processes that create some change in the world. Such events incorporate the qualisign constraint and introduce further constraints entailed by the physical realization of signs as existent happenings. These constraints do not merely concern the need for a sign to be physically perceptible, but also relate to the practicalities of actually realizing and processing signs in communicative contexts. Numerous types of dynamic processing constraints have exercised researchers of the psychology of language for well over a century (Zipf 1935, 1949; Levelt 2013; Levinson 2000; Christiansen and Chater 2008; Cutler 2012). On the speaker side, an *articulatory bottleneck* constrains speakers in language production (Levinson 2000). Certain "absolute physiological constraints on the articulators" result in a "brutally slow" transmission rate of human speech" (28). On the interpreter side, we are constrained in our capacity to process information as it comes in, for example by limits on how many things an interpreter can attend to at once. For all parties there are constraints associated with both long-term and working memory, biases and preferences in reasoning, and the trade-off between minimizing costs and maximizing benefits in processing and action. These limitations foreground the microgenetic frame, the causal-temporal frame of moment-by-moment cognitive and motor processing (Enfield 2014). There are also cognitive constraints of a longer-term kind in the ontogenetic frame. A *learning bottleneck* (Smith, Brighton, and Kirby 2003) places constraints on how, and how quickly, we can learn semiotic systems based on the sinsigns we encounter, and thus limits the kinds of semiotic system that can culturally evolve.

Third, and relatedly, if I am to communicate with you, the signs I use must not just be perceivable and perceived by you, they must also be *interpretable* by you. All semiotic processes involve various types of *ground*, defined as the relation between a sign and its object (i.e., the kind of relation that causes you to take just that sign to stand for just that object: these relations are iconic, indexical, and symbolic). So, to take a sign to stand for an object, you would need to be capable of seeing a similarity between the sign and some object (by an iconic ground), or a relation of physical or causal connection (by an indexical ground), or you should have learned a connection determined by the local-historical conventions of your community (by a symbolic ground). These constraints on interpretation serve as a guide for semiotic agents who must anticipate and exploit their recipients' capacities for perception, attention, and interpretation (Krebs and Dawkins 1984).

We have emphasized the reductive function of the semiotic bottleneck, its ways of subtracting the possibilities for what can serve as a sign. But it can also have an *additive* effect. By means of *collateral effects* (Sidnell and Enfield 2012), the production of a sign may unavoidably introduce other signs.

In relation to constraints on sinsigns (the second point, above), collateral effects arise when certain physical signs can't be realized without also realizing other sinsigns, which were not planned to be part of the message but are then vulnerable to being interpreted as such. As an example, if I use hand gestures while describing a speeding truck's wobbling motion as it lost control along a highway, my hand movements will necessarily have values in an absolute frame of spatial reference, that is, they will be directed toward specific compass directions. In some cultural contexts, this information will be ignored, while in others (e.g., in certain Aboriginal Australian communities) it will be noticed and taken literally, as if I had intended to convey not only that the truck was moving in a certain manner but that it was travelling due north (see Haviland 1993, 1996, 1998; Levinson 2003).[1]

In relation to constraints arising from the capacities and propensities of interpreters (the third point, above), people bring all sorts of interconnected personal, cultural, and linguistic knowledge and capacities to the task of interpreting signs. Anything we say can and will give rise to associations and understandings we may not have intended. Semiotic agents are in the habit of applying ampliative inferential processes to the signs we put in their midst. So, as soon as we have executed a sinsign (or ensemble of sinsigns) for them to interpret in a given move in interaction, we hand over the reins of the semiotic process as it passes through the bottleneck. This means that in any act of signifying, we must anticipate any inferential additives that our interpreters are likely to insert. In this way, the ability to use language not only requires that we can put together phrases and sentences, nor even that we know how to invite inferences that we would want our recipients to make. It also requires that we know how to *avoid* the introduction of unwanted inferential additives to the semiotic processes that we initiate as we present signs to others in context. Indeed, the unceasing endeavor to avoid unintended understandings is arguably the most important mechanism by which the semiotic behavior of community members remains as conservative as it does.[2] As Karl Popper ([1974] 2005, 29) wrote, "it is impossible to speak in such a way that you cannot be misunderstood."

Any form of social engagement presupposes mediation by some set of such signs, within the complex set of constraints that is the semiotic bottleneck. This in turn implies a causal-temporal relation of sign and interpretant, and of both to subsequent signs. That relation may be cast in any number of ways: action and reaction, stimulus and response, question and answer, command and compliance, and so on. All of these point to the "forward-feeding temporal, causal-conditional trajectory of relevance relations" that we call *enchrony* (Enfield 2009, 10; 2013, 29; in press). Enchrony then names a fundamental and necessary frame for all human activity and it is crucial for understanding how intersubjectivity is generated within forms of practical engagement.[3]

The paradigmatic case of an enchronic process is talk-in-interaction (i.e., conversation) between copresent human individuals.[4] This activity is constituted by two partially independent enchronic systems—turn-taking and sequence organization—which together provide for the most elaborate and detailed form of intersubjectivity in the animal world, which in turn provides a foundation for all cultural conceptualization, social organization, and institutional reality. This infrastructure for intersubjectivity emerging at the intersection of two enchronic systems has another key feature. Each move within it is tied to others not only by relations of relevance (every move will be interpreted as relevant to what it is taken to respond to) but also by relations of accountability (agents will be held to account for producing moves that are below a certain threshold of subprehension). Each move can be quizzed, reported, and sanctioned, each of which presupposes the reflexive capacity of language: our ability to use language to talk about language. This insight supports our claim that the enhanced intersubjectivity underlying *all* human forms of engagement—across activities, scales, and units of personhood—is never wholly independent of language.

4.1 From Sawing Together to Talking Together

We have been developing the argument that intersubjectivity is a unique product of engagement by two or more individuals in some form of activity. When Lao monks saw planks together some degree of shared understanding is clearly prerequisite to their participation. They have to recognize one another as possible collaborators in the activity, they have to have some sense that this is a desirable, or perhaps required, activity, and so on. But there's also a kind of understanding that is not a *condition* for, but rather a *consequence* of, their activity. At a most basic level, when one monk pushes the saw, the other monk knows that he is still working toward their common end, and he knows that it is his turn to guide the saw. And when the other monk reaches full extension, the first monk knows that it is now his turn to guide, and so on.

In the case of sawing planks it is easy to see how a basic form of understanding precipitates each move. But what about that understanding is *presupposed by* the activity? This is also a precipitate, a consequence of human engagement, but now on a different temporal scale. For instance, the shared understanding of why sawing planks is a reasonable or valuable thing to do comes from the monks' mutual engagement in the monastic order, which defines merit in terms of good works done in the service of that order (Hanks 1962; Bunnag 1973). Mutual engagement underlies all human understanding at all scales.

Recall our argument from chapter 1: activities of copresent interaction can be arranged along a cline from more to less complex, and the forms of understanding generated through participation mirror this in their internal structural complexity. Thus, the kinds of understanding generated by the activity of sawing are quite simple, those of weaving more complex. In these activities, the internally generated understandings that are both required and possible are shaped, in large part, by the nature of the artifacts involved. One monk knows what the other has done, and anticipates what he will do, through the sensory impressions he receives through his hands. Of course, other things are happening, and other sources of information are available, but the options that the system of sawing provides are strictly constrained and thus the "messages" that can be "sent" and "received" are relatively simple. Weaving involves more scope for selection in formulating any given move and in this allows for more complex forms of understanding of "what we are doing" and "what should happen next." Crucially in weaving there is a distinction between taking a turn (as reed feeder) and what one does in that turn (with options: insert a red or green reed, or insert the thick end versus the thin end first, etc.). This distinction does not exist in the sawing activity, in which there is only one way to take a turn.

Let us now consider the most complex form of interactional mediating structure that we know of: ordinary conversation or copresent talk-in-interaction. In conversation, two distinct forms of enchronic organizational infrastructure—*sequence organization* (Schegloff 1968, 2007a) and *turn-taking* (Sacks, Schegloff, and Jefferson 1974)—define a framework of interaction within which participants engage one another and in relation to which they become accountable according to the normative statuses they occupy. Consider a simple example. If Kim asks Pat a question such as "Where are the scissors?", at the completion of Kim's turn, Pat becomes accountable as the *selected next speaker* (a status created within the framework of turn-taking) and as *the question's recipient* (a status created within the framework of sequence organization). These statuses are built into the architecture of talk-in-interaction. The first status—selected next speaker—normatively obliges Pat to speak. The second status—question recipient—normatively obliges Pat to produce a response to the question, preferably an answer. While often overlapping, these structural organizations and the statuses they confer are conceptually distinct. Let us now unpack these ideas.

4.2 Units in the Organization of Talk-in-Interaction

Key to any understanding of talk-in-interaction is the idea that it unfolds in time (Schegloff 1967; Heritage 1984b; Enfield 2013; see also Bourdieu 1977). But when we think about interactional time or enchrony it is important to distinguish two partially independent unit types that can be conceptualized as endogenous metrics or standards of measure. First, language-mediated interaction unfolds in turns-at-talk and more precisely in *turn constructional units*. These units have a property of *projectability*. A listener can anticipate or predict what the structure of a turn will be while it is still unfolding in time. Participants use turn constructional units to locate or anticipate points of possible completion in a current turn (Sacks, Schegloff, and Jefferson 1974). Participants in interaction are pervasively oriented to these units (see Szczepek Reed and Raymond 2013 for a review of the evidence).

While they are talking in turn constructional units, and *by means of* doing so, participants are at the same time producing actions or moves in a sequence of alternating actions, such as question-answer, greeting-greeting, and request-compliance (Enfield and Sidnell 2017). These sequences of action are linked together by relations of relevance both prospectively and retrospectively (described by Schegloff 1968, 2007a; Schegloff and Sacks 1973). Thus, a current action can reveal how a prior action was understood while at the same time establishing what would be an appropriate next action (see also Heritage 1984b).

An important but underappreciated fact is that these two levels of structure—turn-taking and sequence organization—are organized in partial independence of one another. Whether a person is complaining, inviting, requesting, agreeing, excusing, accepting, granting, or whatever else, the current turn constructional unit will eventually reach a point of possible completion, and when it does, it may convey that transition to a next speaker is relevant.[5] In this way, the turn-taking machinery marches on independently of whatever speech acts are being accomplished in and through its operation. Turn construction plays a crucial role in conveying to the recipient what action is being done (see Enfield and Sidnell 2017b) but the essential design features of the turn-taking (and turn-constructional) system operate independently of whatever else the participants are doing. The turn-taking system is therefore *generic* in that it is not specific to the cultural, social, or even situational particulars of the events in which it operates.[6]

Recognizing this, we can notice that a current utterance displays an understanding of a prior utterance both in terms of its status as a turn *and* in terms of its status as an action (Schegloff 1992; Sidnell 2010). In many cases, a next speaker may have correctly parsed the emerging talk and responded in such a way as to show an accurate understanding that the turn had reached possible completion but may have failed to appreciate what action the first speaker had meant to accomplish by that turn. That is, they can get the turn-taking right but the sequence organization wrong. And the reverse also occurs. A next speaker's beginning to talk can reveal that they misprojected the completion of the current turn even though the action that they produce (or begin to produce) may be well-fitted to the context.

4.3 Taking Turns

The turn-taking system provides for the orderly distribution of opportunities to participate in talk-in-interaction. The system has two components (Sacks, Schegloff, and Jefferson 1974). A *turn constructional component* defines the units out of which a possible turn can be constructed and that, by extension, will allow participants in interaction to anticipate the possible/probable extent and shape of any actual unit and thus to project or predict its completion. This component consists largely of the linguistic structures provided by the language being spoken: the words, grammatical constructions, phonological and prosodic systems, and markers of textual cohesion and interpersonal affiliation. A *turn allocation component* specifies an organized set of practices by which transition from a current speaker to a next speaker is managed. Together these two components and the rules that organize the relation between them provide for an orderly distribution of opportunities to participate and a system in which, as Sacks and colleagues (1974:700) put it, "one speaker at a time" is organizationally primary.

The turn-taking system for conversation ensures that participants work to minimize the amount of overlap in speech (i.e., they avoid having two or more people speaking simultaneously) and, at the same time, minimize the extent of silence between turns in a sequence. Having such imperatives does not entail that all conversation actually proceeds one speaker at a time. Rather, the "one at a time" rule is a normative expectation, thus being at once subliminal, abliminal, and inference-vulnerable: violations of the expectation draw attention and may serve as a basis for inference. In fact, overlaps and gaps between speaking turns are ubiquitous (Jefferson 1973, 1974; Sacks, Schegloff, and Jefferson 1974; Schegloff 2000b), and they can have functional effects arising from the very fact that people perceive them as perturbations.

Overlapping speech is frequent in conversation, but rather than constituting an exception, such cases provide evidence in support of turn-taking rules. When a participant self-selects as the speaker of a next turn, the system dictates that they should begin speaking at (and not later than) a point where the previous speaker's turn is possibly complete and where transition to a new speaker is thus pragmatically relevant. In interactions involving more than two people, multiple participants may target exactly the same point within the unfolding course of talk as a place to begin their own turn. Similarly, a current speaker may reach a point of possible completion but elect to continue speaking. If another participant self-selects, targeting that point as a place to begin their own turn, the result is transitory overlap. So, such points of possible completion within the current course of talk are overlap-vulnerable, this being an effect of a system in good working order.

Consider the following case in which three participants—labeled "Tourist," "Parky," and "Old Man"—are engaged in a casual exchange while sitting on a bench in the park being asked about in line 1. When Tourist asks if the park has changed much, Parky responds immediately at that turn's first point of possible completion with "Oh yes." In asking this question, Tourist has conveyed her own lack of knowledge as well as her sense that Parky and Old Man are relatively more knowledgeable (see Heritage and Raymond 2012). Parky's response—prefaced as it is by "Oh" (Heritage 1984a, 1998)—affirms this assumption. Notice now that when Old Man says "The funfair changed it," he is elaborating and refining Parky's confirmation. The two speakers are thus acting together in an effort to answer Tourist's question. At the first point of possible completion in Old Man's turn (but not before), Parky attempts to begin a turn that will build upon what Old Man has said. Parky twice attempts to begin his turn "That changed it" (before it is eventually produced in full at line 5), yet finds himself in overlap on both occasions (line 4). Notice the split-second timing here, with Parky attempting to come in at just those points where Old Man has reached possible (though obviously not actual) completion of his current turn. In order to come in at just these points, Parky must have anticipated where Old Man would reach possible completion of his in-production turn, projecting from what had so far been said to a point at which it might end.

(1)

01	Tourist:	Has the park changed much?
02	Parky:	Oh yes.
03	Old Man:	The *Fun*fair changed it an [awful lot [didn't it?
04	Parky:	[Th- . . . [That-
05	Parky:	That changed it.

The point can be visualized as shown in figure 4.1.

Such precision timing based on the projection by potential next speakers of the possible completion of a current speaker's talk can be observed in any speech community. Despite occasional claims to the contrary (e.g., Reisman 1974), research has conclusively demonstrated that the system of turn-taking initially described by Sacks, Schegloff, and Jefferson (1974) is a species-specific adaptation to the contingencies of face-to-face human interaction (see Sidnell 2001; Stivers et. al 2009; Levinson 2016). Consider in this respect the following example from a Kri-speaking home in upland Laos.[7]

In this fragment, two Kri men, neighbors and friends, are chatting over tea in the morning.

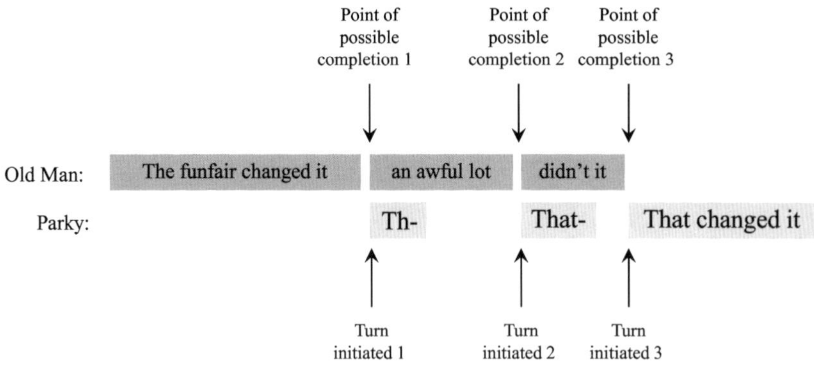

Figure 4.1
Lines 3–5 of example (1), in which Old Man has the floor, and Parky takes Old Man's points of possible completion as places where he could start talking, but on the first two attempts finds himself talking in overlap with Old Man. And so, in accordance with the "one at a time" rule in conversational turn-taking, he abandons his talk in progress. Then on the third attempt he is in the clear and can complete his turn ("That changed it").

(2)

01 Si: *pathềềt muu*
country others
"In their country (the country of the ethnographer)
02 *cààw meengq qa-nìì dơk*
speak language one-this and
one speaks this language and (i.e., Kri, in the recordings)
03 *apoo hanq lôôh meengq qa-qơơk*
soon it come.out language one-other
then it comes out as another language (i.e., when translated)."
04 Ta: *hanq lôôh [mêêngq qaa–*
it come.out language um
"It would come out as the language of, umm—"
05 Si: *[vềềr dêêwq nìì paa vềềr peeng kuu cơng*
return from here they return fix to see
"(When the ethnographer has) returned from here, they return and fix (the language) to see ["fix" = translate into English]

06 Si: *hanq lôôh meengq qa qo·o·k. [meengq cawq qa-nii nii*
it come.out language one-other language we one-this here
"it comes out as another language. This language of ours here (Kri)."

07 Ta: *[meengq-*
language
"The language of . . .

08 Ta: *qò·ò· [lôôh me-*
yeah come.out lang-
"Yeah, it comes out as the lang—"

09 Si: *[pìì nààq thro·o·q*
like that clever
"Like that, so clever."

10 Ta: *lôôh meengq qangkit*
come.out language English
"It comes out in the English language."

11 Si: *qò·ò·*
yeah
"Yeah."

Here Si and Ta are talking about what the ethnographer (Enfield) will do with the recordings he is making in the village. In lines 1–3, Si asserts that in the ethnographer's country the talk they produce will "come out" in another language, that is, it will be translated. At this point Si takes a long (four-second) pause and Ta starts speaking, apparently in an attempt to develop what Si has begun. Si, however, continues on from what he was saying in line 3, substituting, in line 5, *peeng* "fix, alter" for what he earlier referred to as *lôôh* "come out." He then repeats what he has said before with *hanq lôôh meengq qa qo·o·k*, "It comes out as another language." Here he comes to a clear point of prosodic and syntactic completion.[8]

Not surprisingly then, in line 7 Ta begins to talk just as Si reaches this point of possible completion, producing the word *meengq* ("language"). Si, however, continues, also producing the word *meengq* as the start of what is a transition-space repair that replaces the pronoun *hanq* ("it") with *meengq cawq qa-nii nii* ("this language of ours here"). Then at lines 8 and 9, Ta's *qò·ò·* ("Yeah") is treated as a complete response such that the rest of Ta's talk is overlapped by Si's *pìì nààq thro·o·q* ("Like that, so clever").

In both the American park and the upland Lao village, participants in interaction not only parse the talk into formally identifiable units but also project and anticipate the points at which those units will be possibly complete, targeting these as points to begin their own turns at talk. Crucially, participants are *normatively accountable* in relation to this organization. The organization is intersubjectively oriented-to. It thus allows a speaker to find, for instance, that they have been "interrupted" (something that, despite the overlap, does not happen in either of the cases we have considered). This organization also allows participants to perceive that an answer is delayed and to infer from this that, perhaps, the response is not what was expected or preferred (Sacks [1973] 1987; Sidnell 2010). The underlying organization of turn-taking makes even silence meaningful, and this is an effect of the accountability attached to the statuses of question asker and answerer.

The system of turn-taking in conversation is fundamental to human intersubjectivity. It constitutes part of the primary infrastructure that makes language use possible. And in turn, it provides a context in which language can enhance our intersubjectivity. One way in which it does this is through the regulating functions that linguistic details can play within ongoing turns. Such details help listeners to project the structure of turns that are in progress. Another way is through the reflexive capacity that language provides for referring to itself and thereby thematizing the interaction, for example when sanctioning another's speech or holding them to account (more on this below).

The understandings drawn from the turn-taking system are of essentially two types. On the one hand, when people's responses are precision-timed to begin just where a current speaker reaches possible completion, these responses convey participants' attention to the unfolding progress of talk as well as their willingness to contribute to its further development. On the other hand, when people delay their responses, this provides a basis for inference as to the possible reasons for this delay. Perhaps the responding participant is reluctant to deliver the talk that is eventually produced. Perhaps they are preoccupied. Perhaps they don't know how to respond. These are the kinds of information that participation in the turn-taking system contributes in creating the enhanced, deeply layered, and complex intersubjectivity characteristic of talk in interaction.

4.4 Moves in Sequence

In our 2017 book *The Concept of Action* (Enfield and Sidnell 2017a), we examined the theoretical issues implicated in the study of action in interaction and challenged the commonplace and common-sense idea that it involves identifying isolable and readily recognizable action types such as "requests," "complaints," and "invitations." Here we leave those issues aside and focus on the relation between moves in interaction. In semiotic terms, this is a sign-interpretant relation by which every next move is an interpretant of the move that came before it. And crucially, not only that, but every such interpretant (which points back to what just happened) is also a sign (insofar as it points forward to what happens next). This is the essence of the enchronic frame of analysis. Every move in an interaction is a short-lived hinge around which we pivot, pushing ever forward and often changing tack. As we have seen in the examples of sawing and weaving, this relation between moves is the enchronic basis for primary human intersubjectivity. In the case of talk-in-interaction this relation is enhanced by the possibility of linguistic mediation and regulation, and in this enhanced sense it has been described as a relation of *conditional relevance* between moves (see Schegloff 1968; Schegloff and Sacks 1973).

Talk is often organized into sequences of paired moves, known as *adjacency pairs*. For instance, a question is a first-pair part that creates a context within which an answer (as second-pair part) is relevant and expected next. Such paired moves pose two basic problems for analysis (Schegloff 1968,1083). First, "How can we rigorously talk about two items as a sequenced pair of items, rather than as two separate units, one of which happens to follow the other?" and second, "How can we, in a sociologically meaningful and rigorous way, talk about the 'absence' of an item; numerous things are not present at any point in a conversation, yet only some have a relevance that would allow them to be seen as 'absent'?"

To resolve these problems, Schegloff (1968,1083) introduced the concept of conditional relevance:

> By the conditional relevance of one item on another we mean: given the first, the second is expectable; upon its occurrence it can be seen to be a second item to the first; upon its non-occurrence it can be seen to be officially absent—all this provided by the occurrence of the first item.

So, for instance, questions are not always followed by answers. But the conditional relevance that a question establishes will ensure that participants will inspect any talk that follows a question to see if and how it answers that question. A question is a way of imposing rights and duties on someone to provide an answer, in our sense of a normative frame: an answer is abliminal (when it doesn't occur this attracts special attention) and inference-vulnerable (its absence triggers inferences).

What kind of relation is the relation of conditional relevance? It is not a statistical probability and clearly not a categorical imperative. Rather, it is a norm to which people hold one another accountable. In fact, it involves both norms and statuses, as we have defined them. The normative character of conditional relevance is displayed in participants' own conduct in interaction. For example, when a question does not receive an answer, questioners treat the answer as noticeably absent. Questioners' orientation to a missing answer can be seen in three frequently produced types of subsequent conduct: pursuit, inference, and report. The following example (from Drew 1981, 249) provides an illustration of both pursuit and inference. Here a mother, apparently pointing to a clock, asks her child, Roger, what time it is.

(3)

01	Mother:	What's the time- by the clock?
02	Roger:	Uh
03	Mother:	What's the time?
04		(3 seconds silence)
05	Mother:	Now what number's that?
06	Roger:	Number two
07	Mother:	No it's not. What is it?
08	Roger:	It's a one and a nought

After Roger produces something other than an answer at line 2, Mother repeats the question at line 3. Thus, a failure to answer prompts the *pursuit* of a response (see Pomerantz 1984). When this second question is met with three seconds of silence, Mother transforms the question, asking, "Now what number's that?" The first question, "What's the time?" posed a more complex task for the child. The child had to first identify the numbers the hands were pointing to and then use those numbers to calculate the time. With this second question, Mother breaks the task into components. She appears to have simplified the child's task based on an inference from the child's failure to answer, namely, that he was not *able to* answer.

Would-be answerers also orient to missing answers. When a person doesn't provide an answer when an answer is expected, they will often provide instead an account for not answering. A common account for not answering is not knowing. In the following example, Guy has called Jon to see if he'd like to play golf that afternoon. After Jon agrees, Guy asks whether a mutual acquaintance might want to join them.

(4)

01 Guy: Think he'd like to go?

02 Jon: I, uh, I don't know. I'll—I could go by and see.

Here the question posed in line 1 calls for a positive or negative answer even if that is modulated by a marker of epistemic uncertainty (e.g., "I think so, yes," or "No, I doubt it"), given that Jon is being asked to speculate as to some other person's preferences. Jon responds instead with a claim not to know. With this he provides an account for not answering in the normatively preferred way. And he goes further by proposing that he could go by and see, and in this way avail himself of the requested information. With all of this, Jon demonstrates a highly cooperative stance to his engagement in the interaction.

So, the relation between moves is organized by a robust set of norms toward which participants are pervasively oriented. Departure from these norms attracts attention (abliminal) and provides a basis for inference (inference-vulnerable). This underlying structure, powered by the expressive and reflexive properties of language, provides an infrastructure for enhanced human intersubjectivity.

Fixed to this normative chassis is the constant exchange of signs through which participants are able to build up the detailed forms of shared understanding that are at the root of social life. Consider the mundane example of a response to a "How are you?" type of question at the beginning of a telephone call. The superficially simple case in (5) illustrates the rich shared understanding that talk-in-interaction provides for:

(5)

[ring]

01 Deb: Hello?

02	Dick:	Good morning.
03	Deb:	Hi, how are ya?
04	Dick:	Not too bad. How are you?
05	Deb:	I'm fine.
06	Dick:	How'd it go?
07	Deb:	Oh just great, everybody st- still here.
08	Dick:	Oh really?
09	Deb:	Yeah.
10	Dick:	Oh, they stayed. Okay.
11	Deb:	Yeah.

Here, when Dick asks, "How are you?" (in line 4) Deb's response, "I'm fine," is not a mere automatic, conditioned response. It displays a detailed understanding of Dick's turn. First, by starting to talk at this moment (and not earlier), Deb's talk at line 5 shows an understanding that Dick's turn in line 4 was likely finished. Second, by saying "I'm fine"—which here constitutes an answer to Dick's question—Deb demonstrates that she heard the previous turn to be a question. Third, by answering with a description of her personal state, Deb shows that she understands the kind of question it was (i.e., a content question rather than a yes-no question; Fox and Thompson 2010). And fourth, by responding with *fine* rather than *terrible* or *fantastic*, Deb shows an understanding of what this question is doing in this environment, that is, that it is a routine opening inquiry, not to be taken literally (see Sacks 1975; Jefferson 1980; Sidnell 2010).

We can see that this turn-by-turn organization of talk-in-interaction allows for a continuously updated intersubjective understanding shared by the participants and accomplished *en passant* in the course of other activities (Heritage 1984b).[9] Of course, these publicly displayed understandings are provisional and contingent and thus susceptible to being found wanting, problematic, partial, or simply incorrect. But crucially, as we shall soon see, when a first speaker finds the understanding displayed by a second speaker's turn to be inadequate, they have recourse to a mechanism for correcting it.

These are, by now, familiar observations about the role that a next turn plays in displaying understanding, and thus maintaining intersubjectivity, in the enchronic context of interaction. But participants do not produce utterances as simple, unanalyzable chunks. Rather, a single utterance is the product of an interaction between speaker and recipient. This means that *intersubjectivity is established and maintained not only between turns but also within them*. Goodwin and Goodwin (1987) write:

> One very productive strategy for uncovering the interactive organization of talk has focused on ways in which subsequent utterances display an analysis of prior ones (Sacks, Schegloff and Jefferson 1974: 728). However, despite the great power of this methodology and in particular its ability to reveal how participants themselves analyze prior talk in a way relevant to the activities they are engaged in, there are limitations to it. . . . while subsequent utterances can reveal crucial features of the analysis participants are making of prior talk they do not show how participants hear the talk as it is emerging in the first place, what they make of it then, and what consequences this has for their actions, not in a next turn, but within the current turn.

To appreciate this point, we need to acknowledge that many forms of conduct in interaction are *not* turn-organized and thus do not neatly map onto the prototype of enchronic interaction in which the "move"—a bundle of information that serves to convey a social action—is the unit of analysis for identifying the relevant signs and interpretants in a sequence. Running alongside and through the starkly enchronic move-begets-move level of interactional organization, many elements of meaningful conduct can operate either below the level of the turn or above and beyond it.

Turns, as described by Sacks, Schegloff, and Jefferson (1974), are units built up out of talk and, as a result, have certain affordances and impose certain constraints on their use. For example, talk can be impaired by overlap. This fact underlies the basic design feature of the conversational turn-taking system: "For conversation, preservation of 'one party talking at a time' is organizationally primary" (Sacks 2004, 37).

Gesture, by contrast,[10] can be used to display a recipient's understanding within a current turn-at-talk in such a way that the display does not compete with or impair that talk (see Enfield 2009; Goodwin, M. H. 1980; Goodwin and Goodwin 1986; Sidnell 2005). Eye gaze, similarly, can be deployed within or beyond the limits of the turn. When someone is looking directly at a speaker, it will be understood that this person is attending to what the speaker is saying. Conversely, a gaze directed away from a speaker may be treated as evidence that the recipient is not attending, and speakers act in ways that secure a recipient's gaze if it is not directed toward them (see

Goodwin 1979, 1980, 1981). And certain vocal phenomena need not be turn-organized. For instance, laughter can overlay an utterance or be produced simultaneously with an ongoing utterance by a recipient without violating the basic provision for one-at-a-time talk that regulates turn-taking (Sacks, Schegloff, and Jefferson 1974; Jefferson 1979, 1984, 1987; Schegloff 2000a; Glenn 2003). So, a recipient's display of understanding can be produced simultaneously with the speaker's turn-at-talk, meaning that such a display need not be left until the next position. As an example, imagine that a speaker is saying something that is meant to be nonserious, that is, they are joking or being sarcastic. Such moves are frequent but high-stakes, as they risk being understood as meaning the exact opposite to what was intended (i.e., if the hearer interprets the assertion as intended seriously). In this kind of situation, a listener has an incentive to display as soon as possible that they understand the nonserious nature of the move. They may use facial expressions or laughter, for example, to do this in a way that does not intrude into the speaker's turn-space.

There are also practices deployed by speakers in the course of a turn-at-talk that make such responses relevant independently of, and sometimes prior to, utterance completion. An example is the phenomenon of *try-marking* (Sacks and Schegloff [1979] 2007; Enfield 2009a), which works as follows. Speakers often refer to people or things in ways that presuppose that their hearers recognize who or what is being referred to. Suppose someone says "Last night I saw Kim." The use of the first name *Kim* presupposes that the hearer should recognize who is intended by this reference form. But sometimes a speaker isn't fully sure that the hearer will achieve reference. Rather than reformulating the reference to reflect this, for example, by saying something such as "Last night I saw my niece," they may maintain the optimistic first-name reference but add a form of prosodic inflection: the name will be pronounced with the rising intonation usually used for questions. The effect is that the words make a statement while the prosody poses a question. So, when a speaker says, "Yesterday I saw Kim?" this "try-marked" first-name reference ostensibly presumes that the hearer will know who Kim is, but nevertheless invites confirmation from the hearer that the reference is indeed recognized.

In the example just given, the placement of try-marking coincides with the point of the turn's rightful completion. But try-marking may also be used to insert an exchange in the middle of an ongoing turn-at-talk. In the following example, three children are doing crafts together when one of them picks up what he is making and says "Cindy?" (line 1), (meaning something like "I'm going to say something about Cindy, do you know who I mean?") directing this toward the only adult in the room (Eric). By responding "Yes?," in line 1, Eric conveys something like "Yes, what do you want to say about Cindy?"

(6)

01	Ben:	Cindy?
02	Eric:	Yes?
03	Ben:	Has these.
04	Eric:	Has?
05	Ben:	The melter beads.
06	Eric:	Oh ok, did you guys—were you guys with Cindy last year?
07	Carrie:	Yeah.

With line 1 Ben produces try-marking prosody on the first word of what will become a complete utterance. This invites the recipient's recognition of the topic of what is to come. When the recipient confirms that he recognizes who is being referred to in line 2 with "Yes?" Ben then completes the utterance, bringing it to a *first* point of possible completion with "Has these" (in line 3). The example illustrates interaction *within* a turn unit, interaction that provides for the establishment and maintenance of intersubjectivity while the turn is still under way. Specifically, the practice allows Ben to know that the adult is able to recognize the person he refers to as "Cindy" before he proceeds with the rest of his turn. This illustrates the kind of enhanced intersubjectivity that is only possible thanks to language, because it is language that allows Ben to know, at the completion of the word "yes" in line 2, that Eric knows who is meant by "Cindy." (And conversely, Eric now knows that Ben knows that Eric knows who is meant by "Cindy.") All this is established from two words produced as a mere component of a larger utterance.

Let us summarize what we have proposed about the enchronic architecture of intersubjectivity in current and next turns. During the course of a turn-at-talk the speaker monitors the recipient's behavior for evidence of how that turn-at-talk is being understood. Such evidence includes, for instance, their gaze direction, their body orientation, laughter (or lack of it), continuers, assessments, and (especially on the telephone) the audible quality of their breath. Some aspects of a recipient's behavior may suggest a problem of understanding: if someone is not looking at the person who is speaking, it may suggest that they don't regard the talk as addressed to them; if someone isn't laughing, it may suggest they don't understand the talk to be nonserious or joking; if someone fails to say "uh huh" or "yes" at a key moment, it may suggest that they have not recognized who has just been referred to. In such cases, while still in the middle of producing a conversational move, a speaker can redesign their talk on the fly in an attempt to remedy the problem (see Sidnell 2010). Nevertheless, while this sort of to-and-fro is taking place at the sub-move level, each turn will still be normatively required to come to a recognizable point of completion as a unit contribution to the enchronic trajectory of the activity being built out of talk, and in turn, it will be responded to as such a unit.

The result is a highly efficient system by which detailed and complex understandings are displayed in the course of whatever business is otherwise being done and in which problems of understanding can be detected and dealt with more or less immediately.[11]

4.5 Repairing Problems and the Accountability of Talk

In all forms of human interaction, participants encounter troubles of speaking, hearing, and understanding. *Repair* refers to an organized set of interactive practices by which such troubles may be addressed at or near the point where they occur, and may be potentially resolved more or less immediately before there is intolerable divergence of the participants' intersubjective understandings of what is going on in the interaction (Schegloff, Jefferson, and Sacks 1977). The repair organization we will describe here does not constitute *another*, independent piece of the infrastructure for interaction relative to which participants become accountable. Rather, practices of repair (1) operate *over* that infrastructure, and (2) are built up out of structures of turn-taking (composition) and action sequencing (described above).

Although repair is usually defined as a set of practices used to address problems of speaking, hearing, and understanding, in this chapter we are primarily concerned with repair as an instrument of accountability. Here is what we mean by this. Repair targets a particular bit of talk and, in doing so, treats that talk as inadequate: inaudible, incorrect, inappropriate, mispronounced, unintelligible, affected, or whatever else. Repair is never a neutral tool for fixing problems. It is always implicated in the accountability of our behavior in interaction and as such it carries the greatest functional load in upholding the tyranny of accountability in social life. As Sacks once noted, *anything* said in interaction may be targeted for repair.

Practices of repair can be described in terms of (1) *personnel*, that is, *self* (speaker of the trouble source) versus any *other* participant, (2) *positions* relative to the trouble source from which repair is initiated or effected (*same turn, transition space* between turns, *next turn, third turn, third position*, and *fourth position*), and (3) *components* through which it is carried out (*initiation* versus *repair proper*). (See, *inter alia*, Schegloff, Jefferson, and Sacks 1977; Hayashi, Raymond, and Sidnell 2013.) Together, these three sets of distinctions—the personnel, positions, and components of a repair sequence—define a space of possibility for how any repair sequence might emerge or be generated in a stretch of social interaction.

Consider the following case excerpted from a talk show in which Ellen DeGeneres is interviewing Rashida Jones.[12] It begins with DeGeneres raising the topic of Jones's new television show with comedian Amy Poehler, *Parks and Recreation*. DeGeneres initiates the topic by inviting Jones to tell the audience about the show. She then gives the title before concluding the turn with "and you and Amy Poehler how—how great is that," using an interrogative form—*how great is that*—that invites an immediate response from Jones. As we shall see, just what Jones should do in that turn emerges as a source of trouble.

(7)

Ellen: Alright, tell people about this hilarious show.
It's *Parks and Recreation* and you and Amy Poehler how—
How great is that.
Rashida: It's pretty great.
Ellen: mm-hm.
Rashida: It's, uhm, it—I just mean it—ek—
experientially for me it's pretty great (laughing)
Ellen: Yeah, no. And but I mean it's a—
I, ah—know what you meant.
But I say it's really great. The two of you.
Rashida: Yeah.

The talk at line 3 takes the form of a question and Rashida Jones treats it as one by answering "It's pretty great." But by treating "How great is that?" as an information-requesting question, Jones reveals a problematic understanding of DeGeneres's action, which she subsequently goes on to repair at lines 8–9. Specifically, by saying "*I say* it's really great," DeGeneres conveys that her "How great is that?" (at line 3) was not meant as a question but an assertion (meaning something like "It's great, tell me about it").

"I say it's really great" in line 10 of the example is an instance of repair in third *position* (Schegloff 1992). The term "position" here captures the sequential location of the initiation of repair relative to the action it targets as a source of trouble. In first position is the sequence-initiating utterance "How great is that?" The response to it, in second position ("It's pretty great") reveals a problematic understanding of the first-position utterance. This is repaired in third position (the next action in the sequence following the response). Because the person who flags and manages the trouble is the speaker of the first-position trouble-source utterance (when she clarifies that she was saying something, not asking something), this is an instance of *self-initiated* repair (as are all third-position repairs).

Given DeGeneres's initial acceptance of Jones's utterance (by giving a go-ahead with the continuer *mm-hm* in line 5), we might ask what it was in the talk Jones went on to produce that revealed a form of trouble not apparent in the first part. First, we can simply note that Jones evidently struggles in producing the utterance (in line 6) that follows her initial response. Before bringing her utterance to its first possible completion (at "great" in line 7) she restarts this portion of it three times ("It's," "it—" and "I mean it—") and encounters still more trouble in later parts ("ek—"). At these moments, Jones treats aspects of her own in-progress turn as having been problematic in different ways. Because these efforts to initiate repair are launched before she reaches a completion of her utterance and target an aspect of the turn she is currently producing (e.g., the "It's" she begins with), these are instances of same-turn, self-initiated repair. This is the most frequent form of repair in conversation, partly reflecting the preference for self-correction (Schegloff, Jefferson, and Sacks 1977).

Clearly, Jones struggled to articulate this utterance. What made it so difficult? If we examine the utterance that Jones finally completes ("I just mean . . . experientially for me it's pretty great"), we see that it is devoted to revising (or reframing) the initial response she had just completed ("It's pretty great"). She uses the beginning of her utterance ("I just mean") to frame the turn as part of an effort to repair a potential problem of understanding, and completes her turn with a repetition of her prior talk ("it's pretty great"), thus identifying this as the target of the repair. The talk she produces between these elements—"experientially for me"—adjusts or specifies both *what* she is describing as "pretty great" (the *experience* rather than the show) and for *whom* this is so (i.e., herself rather than the audience). Both revisions target ways in which her prior turn could have amounted to a form of self-praise, a kind of action that people are normatively expected to avoid (Pomerantz 1978). In this case, the location of the repair initiation (following a continuer) and the scope of the repair itself (the prior turn) suggest that this is a different species of self-repair than the ones we previously examined. Because the speaker initiates repair *after* both her own prior turn has been brought to a possible completion (which, as we noted above, is the *first* turn in which the trouble could have been dealt with), and her recipient has composed a brief second turn that acknowledges it ("Mm-hm" in line 5), this is an instance of third-turn repair (Schegloff 1997).

This analysis shows how a response ("It's pretty great") initially treated as unproblematic by DeGeneres soon came to be viewed differently by her, and how the third-position repair she launched by virtue of this was consequential for the sequence as a whole. Jones uses third-turn repair to adjust the action import of her initial response, producing an assessment that praises the experience of working with Poehler (defeating any implication that her initial simple assessment was a form of self-praise). But in so doing, Jones evidently invokes an experience that is uniquely her own (working with Poehler on *this* show), thereby limiting the degree to which DeGeneres (or the overhearing audience) can respond meaningfully to it. Moreover, the adjustment Jones makes (in praising that experience) undercuts the degree to which DeGeneres's own prior turn (in line 3) is recognizable as a compliment of her *in particular*. Thus, it is these two issues that DeGeneres addresses in her third-position repair. "But I say it's really great" emphasizes that DeGeneres's initial turn was proposed as a compliment (i.e., something we *say*, not something we *ask*), while "The two of you" clarifies that Jones was one of her intended recipients.

Finally, consider "it's really great. The two of you." at line 10. The resulting construction (specifically the addition of "the two of you") illustrates self-initiated *transition-space repair*. The addition of "the two of you" elaborates the initial "it," heading off any further misunderstanding of what exactly is being assessed (this having been the source of trouble addressed via third-turn repair). Because DeGeneres produced the initial reference, this is an instance of self-initiated repair, and because it is produced in a place where a turn has been brought to a possible completion (and so where another could start, effecting a transition to the next speaker), it is produced in the transition space.

Our analysis of this fragment is meant to show, then, that while repair practices certainly target problems of speaking, hearing, and understanding, in doing so they simultaneously serve as instruments of interactional accountability. Practices of repair identify aspects and components of more or less immediately prior talk that the participants orient to as inadequate, inappropriate, inapposite, or vulnerable to being misunderstood.

Studies of other related practices further ground our claim about the fundamental connection between accountability and intersubjectivity. An example concerns cases of repair that are initiated by means of the phrase "What do you mean?" in English (Raymond and Sidnell 2019). When used on its own, "What do you mean?" conveys that the offending talk has presupposed more common ground than was available to understand what was meant by the utterance. In one example, Person A asks, at the beginning of a phone call, "What happened today?" and Person B responds to this highly underspecified question with "What do you mean?" This leads the first speaker to add more detail: "What happened at work? At Bullock's this evening." Similarly, at the outset of another call, one friend begins with "What is the deal?" to which the other responds "What do you mean?" The first then says "You're not gonna go?" This form of repair conveys (1) that the first speaker has overassumed knowledge held in common (i.e., they have been too vague) and (2) that the second speaker is not accountable for knowing what is being asked about (Raymond and Sidnell 2019).

The phrase *What do you mean* can also be used in combination with a repeat of what someone has just said. In an example from a conversation among children, one child says that a marble has been wasted, and another asks "Why was it wasted?" which leads a third to respond with "What do you mean, 'Why was it wasted?' You should know why it was wasted!" (see Raymond and Sidnell 2019). Note that this reverses the participants' roles in terms of who is accountable for knowing what is meant.

Thus, while it is true that practices of repair are directed at fixing problems of speaking, hearing, or understanding, this is not their only function. There are two obvious and necessary collateral effects of repair initiation:

1. repair momentarily suspends the forward progress of whatever business the participants are engaged in, and
2. repair directs participants' attention to a particular piece of prior talk.

By these means, a speaker can draw her recipient's attention to a just-used word or a just-produced turn, and to the network of presuppositions that ground its use. Seemingly simple devices like "I just mean" (see Rashida Jones in line 6 of the above example) provide always-available means for maintaining intersubjectivity in real time. In subtle but real ways, even this device has the two collateral effects of repair, momentarily slowing progress of the interaction and drawing attention to the talk itself. Such devices would be impossible if language did not have its defining property of self-reflexive reference.

These seemingly low-level practices of error detection and correction might appear trivial in comparison to higher-level aspects of human affairs such as relationships, cultural values, social structures, and historical institutions. But those higher-level matters could never get off the ground without the near-invisible small-scale practices of repair that draw on, and support, the intersubjectivitive foundations of social life. Repair and its collateral effects both create and presuppose the tyranny of accountability in social interaction. Every time we deal with a perceived problem in talk, we are investing resources into keeping things on track: the activity of the interaction itself, the statuses and reputations of the participants, and the relationships at hand.

4.6 Conclusion

Since humans are not telepathic, we depend on semiotic processes for all forms of collaboration, interaction, and social coordination. These processes are subject to a raft of constraints: limits on what is a qualisign for humans (what things can possibly be perceived); practical constraints involved in articulating, learning, and processing sinsigns (what signs are actually produced); and additive inferences that any sinsign may invite, given the knowledge, biases, and inferential capacities of our interlocutors.

Taken together, these constraints comprise a semiotic bottleneck for human communication. While the effects of the bottleneck are far-reaching, across causal-temporal frames and scales, there is one frame in which the rubber hits the road, where the effects of the bottleneck are directly felt and where the consequences are immediately generated. This is the enchronic frame, the frame that constitutes the infrastructure for human interaction and intersubjectivity. In the enchronic frame, human activities expose understandings to public scrutiny and thus to confirmation, contestation, challenge, censure, and coordination. Enchronic processes therefore not only generate and circulate meaning, they provide for a calibration of meaning that is indispensable for the trade-off between the economy needed for efficient agency in real time and the richness needed for meaningful participation in social-relational life.

*

This completes part I. We have established the foundations of basic intersubjectivity in human interaction. And we have defined the enchronic infrastructure for social interaction. This has necessarily incorporated and presupposed a great deal of information about the lexical and grammatical aspects of language, and the enhanced intersubjectivity that these make possible. We now turn the focus, in part II, directly on these features of the "informational" core of language and its implications for enhanced intersubjectivity and human sociality.

II Elements of Language

5 Reference for Shared Intentionality

> Anything which focuses the attention is an indication.
>
> —Peirce [1904] 1998, "What Is a Sign?"

Reference is a form of joint attention and thus an essentially social act. A successful act of reference involves a person attending to something, creating in another person that same line of attention, and getting these lines of attention to converge through a metalevel correspondence (Tomasello 1999; Kockelman 2005).[1] Whereas purely phatic relations of contact are dyadic, acts of reference involve joint attention and thus presuppose a capacity for *triadic* intersubjectivity: that is, a capacity for two agents to share attention to a third entity.

In this chapter, we describe the foundational capacity of reference as a product of joint attention and joint intentionality. In line with our larger argument, we will consider the consequences of language for this capacity, especially insofar as language, as a tool for reference, may be used to refer to itself. These are consequences for social accountability and the enhanced intersubjective understanding that follows. We shall see that metalinguistic reference allows language to become an instrument of its own accountability.

5.1 The Concept of Reference

All reference involves directing the attention of some other person to something. Whether the referent is a hilltop in plain view, a bird's singing, Gottlob Frege, sorrow, the ideas of Augustine or the concept of liberty, making reference requires bringing the recipient's attention in line with that of the speaker. If human cognition is fundamentally *intentional*—in the sense of being about or directed toward something—then reference is a form of *shared* intentionality in which the cognitive focus of two or more people is aligned and jointly focused. In deictic reference, this directing of attention is accomplished by relating an object of reference to some aspect of the event of speaking—the indexical origo—via a ground. So, for instance, when I point to a book and say *This one* in response to the question *Which are you reading?*, my recipient's attention is directed to the book by relating it to my location and specifying the relation as one of relative proximity or immediacy of access (Hanks 1990, 2005).

In this chapter we develop an account of reference that builds from its simplest manifestation in acts of gaze-following. For humans, gaze-following results from a basic propensity to attend to the attention of others. Because copresent others are able to control their own gaze and other visible signs of attention, they can actively manipulate another's attention such that what was a cue becomes a signal (Krebs and Dawkins 1984). Pointing and all other forms of deixis, and indeed all forms of reference, exploit this propensity by actively directing others' attention. Eye gaze in itself does not direct people's attention, but it may be treated by an observer as a cue. This cue may be turned into a signal by the use of certain metacommunicative diacritics, as for example in the practice of lip-pointing in which a facial expression (conveyed by a configuration of lip and or head; Enfield 2009a, chap. 3) indicates that a participants' gaze direction is, at that moment, to be understood as an intentional, communicative signal.

Having established the social and interactional foundations of reference in such scenes of joint attention between copresent persons, we turn to consider the ways in which it both presupposes and elaborates the interactional infrastructure that we have described in previous chapters. We show that deixis, and reference more generally, functions in a context of common ground and background assumptions. When people refer, they typically refer to an already-described world, or to objects within that world that can be located in relation to what is already described. For instance, when, in the third game of the 1932 World Series, Babe Ruth pointed to centerfield, the gesture exploited the recognizable semiotic structuring of the space so as to indicate he intended to hit a home run by sending the ball over the bleachers.

Because reference allows for triadic relations of joint attention, it enables people to coordinate with others around the content of their intentions. This introduces a qualitative leap in the human form of intersubjectivity.

We argue that the momentous shift that reference introduces has another, equally important aspect. It is that linguistic reference—as opposed to mere pointing—can be turned back upon itself such that language becomes both an instrument and an object of accountability. With metalinguistic reference, language users can correct, question, challenge, and otherwise highlight what another has said. This gives human intersubjectivity an inherently reflexive, double-mirrored quality unlike anything else in the animal kingdom. This is what we mean by enhanced intersubjectivity.

5.2 Directing Attention in Deixis

At about nine months of age human infants begin to engage in a suite of joint attentional behaviors such as gaze-following and joint engagement with objects. These behaviors differ markedly from those of younger infants, which are primarily dyadic. At about this age, "infants for the first time begin to 'tune in' to the attention and behavior of adults toward outside entities" (Tomasello 1999, 62). We can think of gaze-following schematically as in figure 5.1.

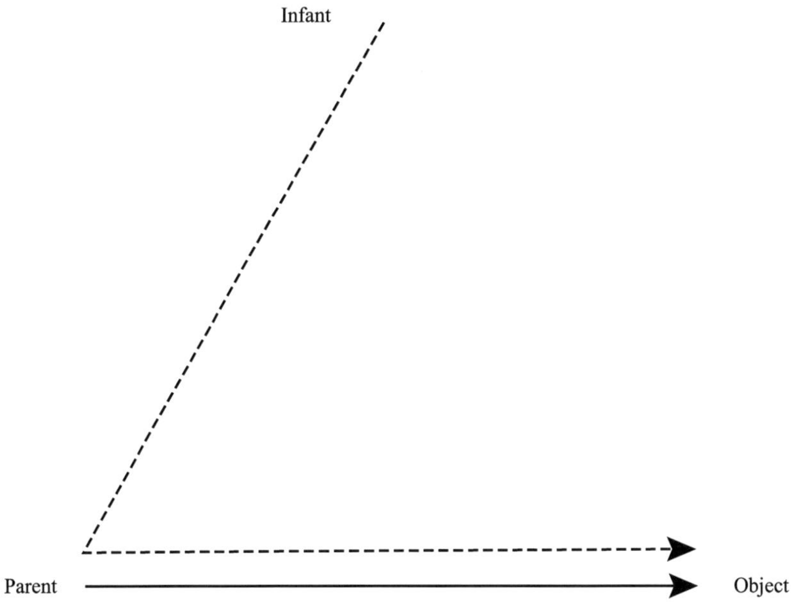

Figure 5.1
Basic structure of gaze-following, with infant and parent.
After Tomasello (1999).

In following the gaze of another, a human infant is attending to that other's attentional state. The infant treats the other's gaze direction as a sign and their own gaze redirection is an interpretant of that sign, bringing the two lines of attention into correspondence (see Kockelman 2005). Gaze-following of this kind may occur independently of whether the other intended their own gaze to function as a communicative signal. In other words, the initial gaze redirection may function to prompt an infant's gaze redirection either as a signal or a cue. As Tomasello notes, it is at around this same age—nine months—that infants also begin to *direct* adult attention using deictic gestures such as pointing, or by holding up objects to show them to others. So ontogenetically, at least, there seems to be a correlation between the emergence of gaze-following and the emergence of deictic pointing and showing.

There is a clear conceptual connection between gaze-following and deictic pointing. Milgram, Bickman, and Berkowitz (1969) showed that gaze-following in adults was sensitive to the character of a stimulus. The study demonstrated that larger crowds of gazing individuals were more likely to induce gaze-following than smaller crowds. The basic, apparently instinctive, propensity of humans to follow the gaze of others is then available for manipulation: altering aspects of the stimulus/sign will make gaze-following by others more likely. This allows us to see the connection between gaze-following and true pointing, one version of which, often referred to as lip-pointing, is done with gaze (it is essentially gaze-pointing). Figure 5.2 shows a San Blas Kuna man lip-pointing (Sherzer 1973):

The term *lip-pointing* "should not be taken to suggest that only the lips are involved. . . . Additional actions of chin-raise/head-lift, gaze direction, and eyebrow raise are usually involved" (Enfield 2001, 186). In the case of lip-pointing among Lao speakers, the vector of pointing is defined by gaze, while jutting out the lips serves a metacommunicative purpose, signaling that the gaze is being used as a point: "the 'vector' of lip-pointing is in fact defined by gaze, and the lip-pointing action itself (like other kinds of 'pointing' involving the head area) is a 'gaze-switch,' i.e., it indicates that the speaker is now pointing out something with his or her gaze" (Enfield 2001, 185).[2] The example of lip-pointing illustrates how humans can accomplish intentional reference (i.e., nonnatural meaning in Grice's 1957 sense) through small manipulations of naturally meaningful behaviors (gaze-direction), thus exploiting the human propensity to follow another's gaze. The introduction of a metacommunicative overlay (chin/lip/head

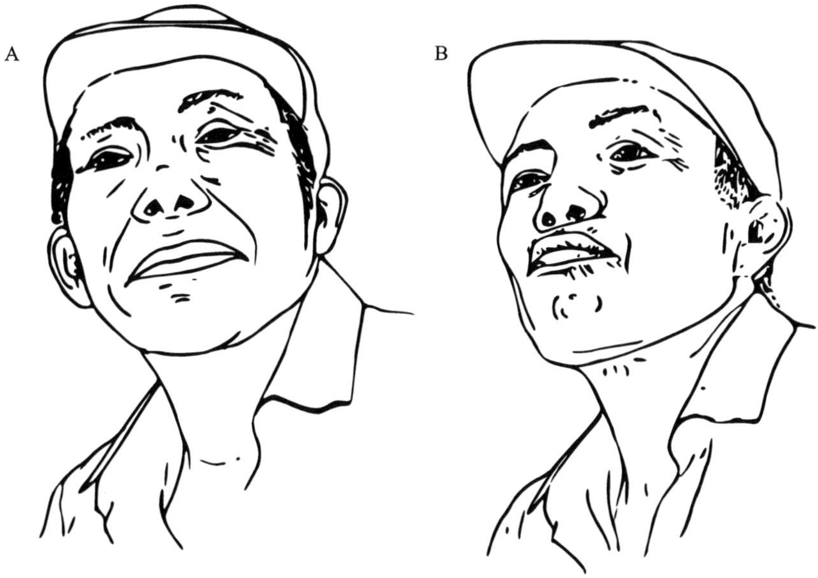

Figure 5.2
"The pointed lip gesture among the Cuna Indians of San Blas, Panama."
After Sherzer (1973, 117).

Table 5.1
Current and projected focus of attention in deixis

Gaze-direction	Lip-pointing	Finger-pointing
CFA only	CFA = PFA	CFA = PFA or CFA ≠ PFA

Key: CFA = current focus of attention, PFA = proposed focus of attention

movement) on gaze direction transforms a cue into something another can recognize as a true intentional signal, interpretable as "He's referring to that thing/person/area over there."

Finger-pointing has a distinctive feature relative to the forms of primitive deixis so far described. In finger-pointing it is possible to separate the speaker's current direction of gaze (typically taken as an index of their current focus of attention) from the focus that they are proposing *for a recipient*. Table 5.1 shows the basic elements and their combinatorial possibilities:

We illustrate the independence of the direction of pointing and the direction of the person's attention (or at least their direction of eye gaze) with images from a videorecorded interaction among speakers of Bequia creole, shown in figure 5.3. In the first frame, Viv (in the foreground) is telling Baga (in the background) about a man she thinks he might know but whose name he does not recognize. When the description given allows Baga to identify the person Viv is talking about, he points up the hill to his right (second frame). Notice that when Baga initially points, his own gaze is directed to the place he is indicating with his finger (so that CFA = PFA). In the third frame, he maintains the pointing gesture but now gazes toward Viv, apparently to check whether his reference has been successful—checking, that is, on his recipient's focus of attention (so that CFA ≠ PFA). He finds Viv pointing to the same place and the two engage in a moment of mutual gaze (last frame). Here then we can see, in the visible behavior of the participants, how reference involves joint attention such that two people are not only publicly

Figure 5.3
Finger-pointing, Bequia, St. Vincent. Stills from video by Jack Sidnell.

projecting attention to the same referent, but are also mutually aware of the current alignment, and thus sharedness, of their two lines of attention.

The possibility of separating a speaker's/gesturer's directing signal from their gaze is important because a joint attentional frame crucially involves the speaker monitoring the recipient's attention to some third object (Carpenter, Nagell, and Tomasello 1988; Liszkowski et al. 2004; Tomasello 1999; Tomasello et al. 2005).[3] This monitoring of the recipient transforms shared attention to a third entity—which may simply be a parallel focus rather than an intersubjectively *shared* focus—into true joint attention, the basic form of shared intentionality (see also Searle 2010; Gilbert 1992).[4]

We can see many of the basic features of deictic reference in another form of behavior among infants, namely, showing (Tomasello 1999; Clark 2003; Kidwell and Zimmerman 2006). In a showing sequence, a young child will approach another person (typically an adult) with an outstretched arm and an object in hand (see figure 5.4); the other might produce a response that

Figure 5.4
Human infant showing object to cameraperson. Photo by N. J. Enfield.

identifies the object ("A lego piece!"), expresses a social-relational feature of the object ("That's your lego!"), or appreciates it in some way ("Oh wow, what a cool lego piece!"). The showing child then withdraws the object and/or moves out of the recipient's line of vision, either returning to the activity she was engaged in before the showing or initiating some new activity.

Such acts of showing are one of the most basic forms that exhibit the triadic, joint attentional interaction configuration that constitutes the foundation of reference in all its forms (see Tomasello 1999, 2003). In both showing and pointing, the first person's action results in the second person's attention being directed toward some object. The key difference is that in pointing, the other's attention is made to move toward the current location of a thing, while in showing, a thing is moved into the current line of the other's attention (Clark 2003).

Acts of showing are a developmentally early form of presentational or demonstrative deixis akin to adult uses of French *voilà* or English *look at this*. Instructional activities build upon the human propensity for attending to the attention of another. Showing plays an important role in their organization. Rembrandt's *The Anatomy Lesson of Dr. Nicolaes Tulp* provides a stunning illustration (figure 5.5).

Here Dr. Tulp is presenting a part of the cadaver for the consideration of his students, some of whom look attentively. Through such acts of showing, novices are socialized into new ways of seeing the world around them, ways of seeing that are appropriate to certain statuses or roles (see Goodwin 1994; Kockelman 2007b). In acts of showing, we see not only the roots of reference in human action but the interactional foundations of human teaching and learning, as well as the transmission of knowledge across generations and thus, ultimately, of culture (see Tomasello 1999).

5.3 Deixis and Common Ground

We began with the simplest kinds of joint-attentional scenes, the kinds that allow a nine-month-old to get started on a long journey of socialization (see Brown and Gaskins 2014). It is a yearslong path of countless moments of joint attention, countless instances of learning and guidance, and of gradual convergence in knowledge and stance with elders and peers, first through simple gestures and shared participation frames, and soon within the increasingly rich matrices of language, kinship, ritual, livelihoods, and

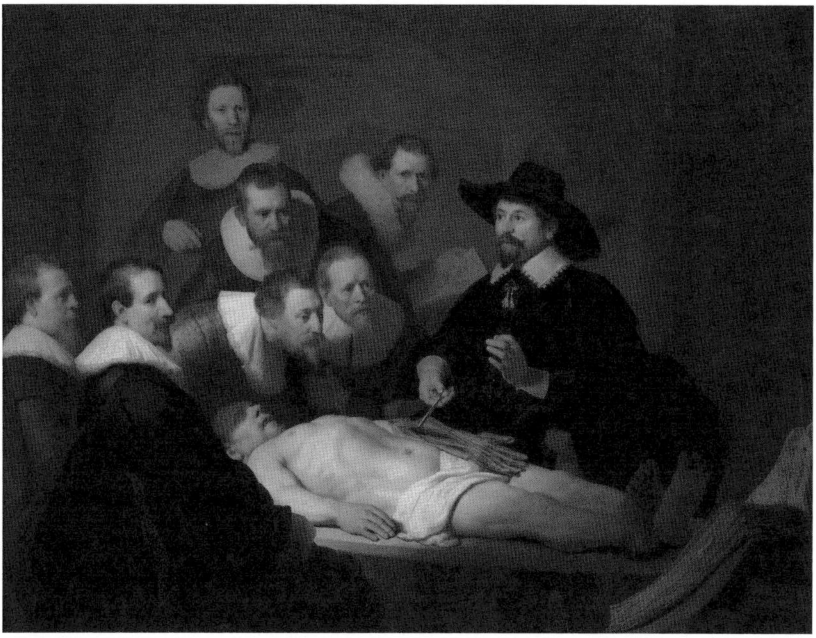

Figure 5.5
The Anatomy Lesson of Dr. Nicolaes Tulp, Rembrandt Harmenszoon van Rijn, 1632. Image is in the public domain.

material culture. These aspects of the socio-cultural world form the basis of a community's common ground, and thus are naturally caught up in the elements of demonstrative systems, dependent as they are on whatever sources of mutual salience happen to be at hand. Much previous work on deixis has been a search for the right gloss of each deictic form's meaning. But deictic terms like demonstratives (*this, that, yon*) and spatial deictics (*here, there, yonder*) resist this, instead requiring interpreters to rely heavily on context in figuring out their reference on any given occasion. The key to interpreting deictic expressions is the common ground between interlocutors (Clark 1996; cf. Hanks 2006). The use of demonstratives in Lao-language interactions in grocery market settings shows how rapid changes in common ground arising from fluidly evolving participation frames can affect the selection of demonstratives ("this" versus "that") for picking out referents that are all proximate and in common view (Enfield 2003). In other contexts, we see how common ground of a more enduring kind—cultural common ground (Clark 1996)—also affects the selection and interpretation of deictic expressions.

Consider the following case of interaction among young children in a rural Guyanese village. Many villagers live on strips of land belonging to their patrilocal group, each strip containing the houses of two or more nuclear families. The houses are on stilts and the downstairs area, termed the bottomhouse, is the setting for much of everyday domestic life. In the fragment of interaction we present here, siblings Kavita, Kumar, and Baby have been playing in the bottomhouse of their family home when Mando, their distant cousin who was being watched by their grandmother at her house in the same yard, runs down the path and enters the area where the children are playing.

(1) (Guyanese Creole)

01	Baby:	*EE GYAL (muv fram deer)*
		"Hey girl, move from there,
02		*WA YU KOM HIA FA?*
		what have you come here for?
03		*Yu wan mi kyeer yu bak a Linda dem?*
		Shall I carry you back to Linda's house?"
04	Mando:	*ya*
		"Yes."
05	Baby:	*wel guwan yu wee*
		"Well go—on your way."
06	Kavita:	*tel-kos shi*
		"Tell—curse her."
07	Kumar:	*riimuv from deer. riimuv from deer Mandoo*
		"Move from there, move from there, Mando."

When Mando approaches the group, her cousin Baby yells at her, asking "What have you come here for?" She follows up with "Shall I carry you back to Linda's house?" thus treating Mando's entering of the immediate area as an accountable action and demanding an explanation. (Linda lives in the next yard over in a house belonging to distant kin and nonmembers of the patrilocal group.) The case shows how a deictic expression such as "here" maps directly onto already described, semiotically defined space.

Baby's "here" refers to the bottomhouse and to that part of the otherwise communal yard that is the property of her own immediate family. In a case like this (recall Babe Ruth pointing over the bleachers in the 1932 World Series), speakers (or gesturers) do not refer to blank space but to an already described, highly structured immediate environment (Goodwin 2003).

Another example involves speakers of Kri. In the Kri-speaking community of Mrkaa Village in upland central Laos, the physical layout of the house (figure 5.6) is a diagram of social-relational asymmetries, on two axes (figure 5.7; see Enfield 2009). Running laterally across the house is an in-out axis, where "in" maps onto "private, family, women, children, storage room, food preparation" and "out" maps onto "public, nonkin, men, adults, guests, drinking, public ritual." Orthogonal to this is an axis that runs from what we would call in English the "front" of the house, where one enters, to the "back" of the house. In Kri, this is referred to as a below-above axis, where "below" maps onto socially lower ranks, and "above" onto socially higher ranks, where relative "height" is determined primarily by relative age, often attenuated by classificatory kinship.

Figure 5.6
Kri house, Mrkaa Village, upland central Laos. Photo by N. J. Enfield.

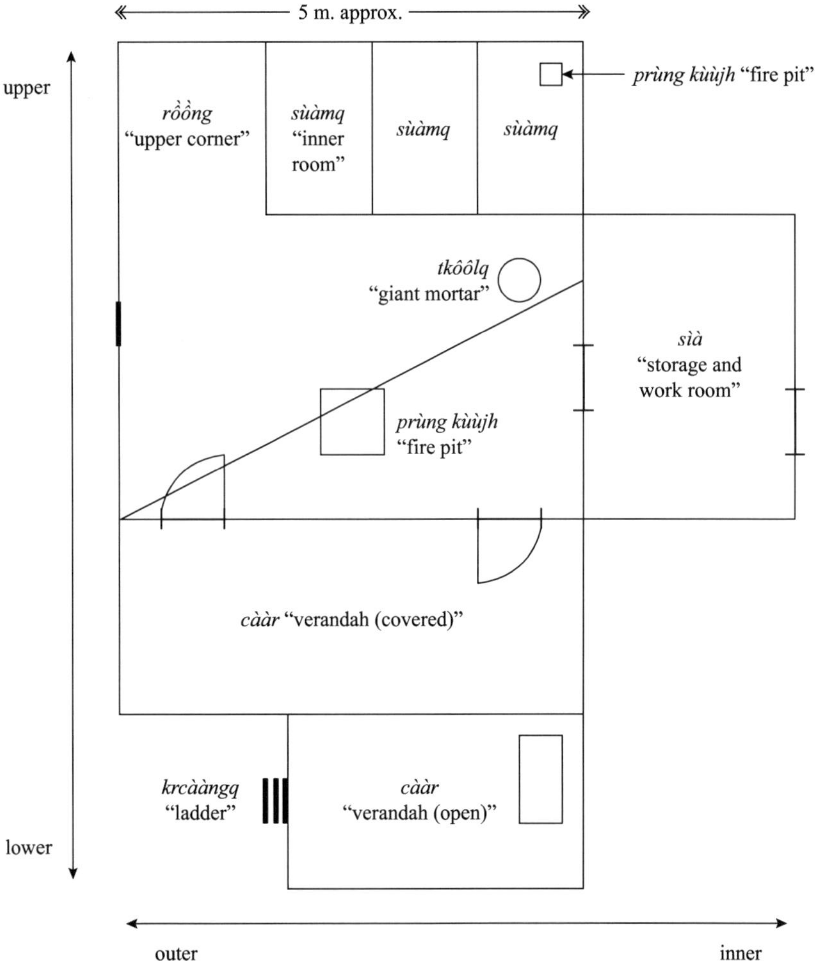

Figure 5.7
Kri house floor plan.

The Kri house is therefore conceptualized spatially as a microcosm of the larger geographical environment, as coded in the demonstrative system. This has typologically familiar "proximal" and "distal" forms (*niiq* "this" and *naaq* "that"), and has three further forms in addition: *lêêh* ("the one up/above/upstream/uphill"), *cồồh* ("the one down/below/downstream/downhill"), and *seeh* ("the one across," that is, away but neither up nor down). While the house floor is normally perfectly level, the "up/down/across" scheme is

Figure 5.8
Kri speakers in interaction, Mrkaa Village, Laos. Still image from video by N. J. Enfield.

nevertheless readily mapped onto it, thanks to its diagrammatic relation to the social-cultural dimensions represented as in-out (family versus nonfamily) and up-down (senior versus junior). Now consider an example from a videorecorded interaction between a group of Kri-speaking women sitting on a front verandah, in which this socioculturally motivated mapping provides the solution for a simple referential problem of locating an object. Figure 5.8 shows a still image from the video recording.

The scene is in the house of Et, the elderly woman at the right of frame. We focus on an exchange between her and Nuan, the young woman second from left, visible in the door frame, who does not live in this house.

(2) (Kri)

01 Nuan: *piin sulaaq*

give leaf

"Pass some leaf."

02 Et: *sulaaq quu kuloong lêêh, sulaaq, quu khraa seeh*

leaf at inside up there leaf at store across there

"The leaf is inside, up there, the leaf in the store."

03 *môôc cariit hanq*

one backpack it

"(There's) a (whole) backpack."

04 [Nuan walks inside]

In line 1, Nuan makes a request to be given some "leaf" (actually, corn husk) with which she can roll a cigarette. In Et's reply, she uses a complex combination of referential expressions to inform Nuan of the location of the "leaf" so that she can go and get some herself. First, an intrinsic spatial reference (*kuloong* "inside") is combined with the "up" demonstrative *lêêh*, in alignment with the up-down axis of the house. From their perspective sitting on the verandah (the "lowest" part of the house), the inside area of the house is "up," and, accordingly, this is coded in the demonstrative chosen. Et then narrows in further on the spatial location. Where they are currently sitting is the "outer" side of the house, and the "leaf" in question is located inside the *khraa* ("storage room") at the "innermost" side of the house. Once one has entered the house going "up" from where the speakers are sitting, one would then have to go "across" (away, but neither up nor down). This is specified with the demonstrative *seeh* ("the one across there").

This example illustrates one way in which the interpretation of demonstratives depends crucially on shared background knowledge—the description of the world that members of a community collectively take for granted—as relevant to the context of speaking. In the case of Kri, the selection and interpretation of demonstratives draws directly on a conventional mapping of the socio-cultural domain of kinship and other personal relations onto the two-dimensional spatial array of the house floor plan.

5.4 Language as the Object and Instrument of Reference

We've seen that reference can be achieved without language, by pointing and other nonverbal means of directing attention. But such gestural practices have their limits insofar as there are kinds of objects that are difficult, if not impossible, to point to. With language as the instrument of reference, a wider swath of the phenomenal world can be indicated and made the object of joint attention than would be possible with gestures alone. A speaker can refer to a stretch of time, a thought or feeling, an abstract quality like *justice*, and even to language itself. With language as instrument, language becomes a possible object of reference.

Metalinguistic reference is pervasive in talk and essential to its own internal organization (see Deacon 1997; Schegloff 1989, 1991; Taylor 2000; Duncker 2019).[5] The reflexive capacity of language has arguably transformed human sociality and is fundamental to the enhanced intersubjectivity that underlies modern human relationships and institutions.

The unique properties of language make it a powerful and precise instrument for intersubjective accountability, as scholars from a range of disciplines have noted. With language we can pick out and thematize any aspect of the phenomenal world that we can perceive, and we can draw others' attention to such features, demand explanations, and so on. To this fundamental observation we are adding another: language also has unique properties as an *object of accountability*. Within the stream of linguistic conduct, one can identify for scrutiny a word, a pronunciation, a phrase, a clause, a larger grammatical construction, a pragmatic or contextual presupposition, and so on. Especially in contexts of socialization, people are constantly offering opinions on which words or expressions are good or bad, suitable or unsuitable, effective or ineffective. No other form of human conduct presents itself in such a way.

Suppose you wanted to identify some aspect of the way a person got up from their chair during a meeting. Yes, this can be described, and if needed it can be described in some detail. But if you want to identify or describe what a person has *said*, the conduct achieved through the use of language can be more efficiently segmented, identified, and so made an object of reference. And various aspects of language function can be highlighted, and hence either challenged or repaired, from referential extension to semantic denotation, from contextualized meaning to pragmatic inference and implication.[6]

We will consider four distinct forms of metalinguistic reference: indexing, repeating, framing, and redoing. In the first two, speakers draw attention to a previous instance of language use (usually immediately previous). Insofar as these practices involve establishing joint attention to an instance of language use, they are acts of reference. In the second two types, framing and redoing, speakers draw attention to previous acts of *reference*. Thus, these are not only metalinguistic acts of reference but metareferential acts of reference.

5.4.1 Indexing as a Tool of Metalinguistic Reference

The simplest forms of metalinguistic reference are akin to pointing insofar as they involve the use of "pure" indexes to merely draw attention to a previous instance of language use. These metalinguistic referential indices are a type of *other-initiation of repair*, variously called "open-class repair initiations" (Drew 1997) or "open-request repair initiations" (Dingemanse and Enfield 2015). Cross-linguistically they are of two types (though we note that there are other specialized ways of doing open-class repair initiation; see Sidnell 2010). One type is an interjection (*Huh?*), which has been found to exhibit great uniformity cross-linguistically (see Enfield et al. 2013; Dingemanse et al. 2013; Dingemanse et al. 2015). The other type is a question word or question phrase in some languages (such as Vietnamese, see below). Consider a simple case from English.

(3)

01 Jon: Well, I'm sure we can get on at San Juan Hills.
02 That's a nice course, I only played it once.
03 Guy: Uh huh?
04 Guy: It's not too bad.
05 Jon: Huh?
06 Guy: It's not too bad.
07 Jon: No.

Jon and Guy are trying to decide where to play golf later in the day and Jon suggests San Juan Hills in line 1. After Jon offers an evaluation of the course as "nice" and also recommends it for its novelty (to him), Guy responds first with "Uh huh?" and then with "It's not too bad," hardly a ringing endorsement. This is responded to by Jon at line 5 with the interjection "Huh?" Guy then repeats what he has just said (line 6). The interjection *Huh?* functions by drawing Guy's attention to his just previous utterance, that is, it is an index (or "indication" in Peirce's early terminology).

Here is an example from a conversation in Lao:

(4) (Lao)

01 Loi: *nòòj4 bòò1 mii2 sùak4 vaa3 nòòj4*

Noi not have rope question Noi

"Noi, don't you have any rope, Noi?"

02 Noi: aa? [hã]

03 Loi: *bòò1 mii2 sùak4 vaa3*

not have rope question

"Don't you have any rope?"

The example is structurally identical to lines 4–6 in (3). Noi responds to the initial utterance in line 1 with an interjection token (*aa?*), and this elicits from the initial speaker a repetition of the first utterance (though without the recipient-selecting person reference—the name "Noy"—as it is now no longer needed in this redoing of the utterance, given that the person's participation is already secured).

A final example is from Vietnamese. Thanh and Phuong have been talking about a time that they went together to sing karaoke in Ho Chi Minh City. Thanh asks Phuong to guess how much it cost and, after some talk in which Phuong indicates that Thanh already told her how much it was, she produces the talk in line 1.

(5) (Vietnamese)

01 Phuong: *Ở đấy tám mư'ơi nghìn một tiếng đúng không*

at there eight ten thousand one hour correct no?

"It is eighty thousand per hour there, right?"

02 Thanh: *Hả*

"Huh?"

03 Phuong: *Tám mư'ơi nghìn một tiếng đúng không*

eight ten thousand one hour, correct no?

"Eighty thousand per hour, right?"

04 Thanh: *Ừ'.*

"Yes."

The sequence in lines 1–3 is structurally identical to those in examples (3) and (4). To an initial utterance the addressee responds with an interjection token (here something like *Hả*) and this elicits a repetition. Our focus here is on the metalinguistic referential function of the interjection token. It functions like a verbal point that draws interlocutors' attention to just-prior talk.[7]

5.4.2 Repetition as a Tool of Metalinguistic Reference

Next-turn repetition is another metalinguistic device that draws attention to the just-prior utterance, thus serving as a form of indexical reference to language. Repetition is never merely or solely repetition. Any repetition necessarily conveys something of the speaker's stance toward the previous utterance. So, even if a subsequent speaker repeats a previous utterance and attempts to preserve its exact intonation, prosody, and stress patterns as well as its exact wording, this can be heard as mimicry or mockery, thus conveying the speaker's stance toward what is repeated (Sidnell 2009). In next-turn repetition of the kind we are concerned with here, a speaker will typically repeat some or all of the segmental structure of what the other person just said, but they will use different suprasegmental features of prosody. These features of prosody may indicate the function being performed by repeating the utterance: for example, conveying surprise or disbelief, suggesting error, or expressing amusement (see Rossi 2020). Our focus here is on the referential function of such repeats. Consider some examples.

In (6), Ben has called to talk to Albie and reaches Albie's wife, Cassie. When Cassie tells Ben that Albie is outside, Ben repeats "He's outside" with rising intonation:

(6)

01	Cassie:	He's outside.
02	Ben:	He's outside?
03	Cassie:	Yeah.
04	Ben:	Can you fetch him?
05	Cassie:	Okay, just a minute.

Albie comes to the phone and he and Ben make plans for a party at Albie's that evening. In (7), at line 1, Ben asks Albie what he plans to drink, and when he does not receive an immediate response he elaborates the question with, "You mixing drinks tonight or are you drinking out of a can or what are you doing?" Albie responds with a repeat of the last part of Ben's talk: "What am I doing?" (with the pronoun adjusted from "you" to "I"). This elicits a confirmation from Ben in line 5 (i.e., confirming that this is indeed what Ben wanted to say at the end of the utterance in line 3).

(7)

01	Ben:	What are you drinking?
02		You mixing drinks tonight or you drinking out of
03		a can or what are you doing?
04	Albie:	Uhhh. What am I doing?
05	Ben:	Yeah.
06	Albie:	Oh shit, I don't know. We got some m—some small
07		bit of that white whiskey left. I think I'll drink
08		that and or some vodka we got . . .

Repetition is a metalinguistic tool for repair-initiation not only in English but in a range of highly diverse languages. Here are examples from Russian, Siwu (spoken in Ghana), and Yélî Dnye (spoken in Papua New Guinea):

(8) Russian (Baranova 2015, 568–569)

01	Ana:	*Gena, gavarit, minia zavut.*
		Gena say-he me call-he
		"My name is Gena," he said.
02	Brit:	*Gena?*
		Gena?
03	Ana:	*Da*
		yes

(9) Siwu (Dingemanse 2015, 235)

01	Ogbe:	*Kófi. Kòfi koso ɛbɛrɛ be ũ*
		Kofi Kofi young or what them
		"Kofi. Kofi junior or whoever"
02	Kuma:	*Kòfi koso:*
		Kofi young
		"Kofi junior?"
03	Ogbe:	[head nod]

(10) Yélî Dnye (Levinson 2015, 393)

01 A: *u kpâm mye Y:oonkîgh:ê pyââ*

his wife also Y:oonkîgh:ê women

"his wife's (female) relatives are also at Y."

02 B: *mye Y:oonkîgh:ê pyââ?*

also Y:oonkîgh:ê. women

"the relatives are also at Y?"

03 A: *nyââ*

"Yes"

Consider also a case from Vietnamese (Sidnell, Trần, and Vũ 2020). The participants, who work together at a health insurance company, are talking about a time that Tiến hosted a gathering at his house, which is some distance from Hanoi (where they work). Mai is explaining, in lines 1–2, that she was busy and couldn't come. By gazing at Tiến while she says this, Mai indicates that she is addressing him. However, it is Lệ who is most active in taking up Mai's talk. In line 3, Lệ repeats *Chị không sang đư'ợc* ("You [elder sister] didn't get to come"). Mai then shifts her gaze to Lệ and responds to the repeat with a brief head nod (line 4) before the talk continues.

(11) (Vietnamese)

01 Mai: [Addressing Tiến]

Hôm sang nhà Tiến chị bận cái-gì-này

day come house Tiến elder-sister busy whatever,

"The day that you had people over, I was busy and whatever

02 *nên chị không sang đư'ợc*

so elder-sister not come get

so didn't get to come."

03 Lệ: *Chị không sang đư'ợc*

elder-sister not come get,

"You didn't get to come."

04 [M head nod toward T]

05 Lệ: *Mhm*

06 Ha: *Em mời lần nữa đi cho*

younger-sibling invite time again go give

"Invite us sisters one more time so

07 Ha: *các chị sang*

group elder-sister come,

we can visit."

We noted above that the particular way in which a repeat is produced inevitably conveys something about the stance that the repeat speaker takes up in relation to the repeated talk. Consider example 12 in which friends Jim and Al are talking about Jim's attempt to secure tickets for an upcoming concert. When Jim reports that the name of the person who helped him was "Smitty," Al repeats this with laughter intermingled with the production of the name. This clearly suggests that Al has found something remarkable, surprising, and perhaps amusing in the name. Jim, however, merely confirms, repeating the name but not taking up the invitation to laugh that Al has issued with his own repeat (see Jefferson 1979). Al first registers the confirmation (in line 4) and after Jim continues with what he was saying in line 5, Al begins a bout of overlapping laughter and subsequently repeats the name "Smitty" again, now with the even more pronounced interspersed laugh tokens. This now elicits the kind of response Al was initially after with Jim saying "Doesn't really sound like someone's real name, does it?" We reproduce the transcript of this example with more of the production details included, as these are central to the case we want to make about it.

(12) Smitty—Concert Tickets

01 Jim: The guy that helped me, his name was Smitty.

02 Al: S:mi(h)::tt(h)y? [h = interspersed with laughter]

03 Jim: Smitty.

04 Al: Uh huh.

05 Jim: So I'm wr—writing [his name down.

06 Al: [*.hhh* *huh huh*

07 Al: S:mi(h)::tt(h)y hhh .hhhh [hh (h = laughter)

08 Jim: [Doesn't really sound like

09 someone's real name, does it.

10 Al: No, it doesn't. Maybe he was bullshitting you.

Cases like this show how repetitions can clearly indicate a repeat speaker's stance toward the repeated talk (as surprising, ridiculous, etc.). The stance so displayed also indicates what kind of speech act the repetition is creating (see Rossi 2020 on action). Action is laminated on the metalinguistic reference done by repetition. Such repeats can be used in a variety of ways and can invoke a range of accountabilities. In the case above, Al uses the repeat to hold Jim accountable for mentioning what he (and eventually Jim) takes to be an unusual name, one that suggests that the person who originally gave it was "bullshitting." A similar case from Lao is given in (13). Here, when the first speaker reports someone paid 100 million, this is repeated—suggesting incredulity—and subsequently invites confirmation.

(13) (Lao) Enfield (2015, 138)

01 Ek: *lòòj4 laan4*

100-million

"[They paid] a hundred million."

02 Thaa: *lòòj4 laan4 phun4 na*

100-million yonder

"A HUNDRED MILLION?!"

03 Ek: *qee5*

"Yeah."

The common feature across all these examples is that a repeat in a subsequent turn is used to draw attention to, and in this sense to refer to, a just-previous utterance. Obviously, such a practice can be used to achieve a wide variety of interactional outcomes. This is an example of the way language serves as an instrument for its own accountability (see Rossi 2020; Jefferson 1972).

5.4.3 Framing as a Tool of Metareferential Reference

Consider now cases in which the recipient of an utterance treats a referential expression in that utterance as problematic. Where the initial referential expression is a lexical noun—such as *plane* in example 14—the recipient can indicate a problem by repeating the noun, prefaced by a question word (e.g., what, which):

(14)

01 Chris: And all the people ran onto that plane. Right?
02 Andy: What plane?
03 Chris: That plane right there. Or this plane. Yeah, 'cause there—
04 Andy: They ran onto that.

In (14), children are constructing a joint narrative around planes made from Lego pieces. At line 1, Chris proposes, "And all the people ran onto that plane. Right?" His companion, Andy, is unable to determine which of two planes he means and responds with "What plane?" The repetition of the noun "plane" serves to frame what is treated as the problematic term ("that"), which is replaced with a question word. This type of strategy is well-documented across genealogically unrelated and typologically diverse languages (see, for example, Dingemanse, Blythe, and Dirksmeyer 2014).

When the original referring expression is a pronominal, we find a slightly different pattern. Consider (15):

(15)

01 Anne: [looks at door] Maybe Rose, maybe you can move it.
02 Rose: Move what?
03 Anne: Move that thing that's in the lock yo—in the door.
04 Rose: Okay.

In this example, Rose has been trying to close the door to the room in which the three children are playing. One of the children has left their cardigan hanging on the door handle and this is preventing the door from closing. Anne, speaking to Rose, suggests that she could move it. Rose, not able to discern what is being referred to, remarks "Move what?" Again, the repetition of a just-used word ("move") frames the problematic term ("it"), which is replaced by a question word.

A similar example is shown in (16) in which two children—Matt and Josh—are facing two chairs that have been set side by side. Matt, noticing a difference in height between the two (identical-looking) chairs remarks, "Wonder why this is higher?" Although Matt and Josh appear to be jointly attending to the same set of objects, this utterance has the character of an "out-loud" remark and is not obviously addressed to Josh. At line 2, Matt announces a just-discovered solution to the puzzle: "'Cause it's made higher." Josh now repeats part of Matt's utterance while substituting a question word ("what") for the pronominal "it," in this way drawing attention to—effectively referring to—Matt's use of the word *it*. Notice that he does not refer to what Matt referred to with "it"—this is precisely what is in question—rather, "what" refers to the expression used in Matt's act of referring. This is metareferential reference.

(16)

01	Matt:	I have the higher place [looking at chairs].
02		Wonder why this is higher? 'Cause it's made higher. Ah . . .
03	Josh:	What's made higher.
04	Matt:	[taps the chair top]
05		See, 'cause look. I can't even get up 'cause it's too high.

Similar types of reference repair are found across all languages and are in various ways adapted to the grammatical particularities of the language in question (Sidnell 2007; Egbert, Golato, and Robinson 2009; Hayashi and Hayano 2013; Dingemanse, Blythe, and Dirksmeyer 2014). In Vietnamese, reference is frequently attempted with null or zero forms, which may resist resolution. In (17), in line 1, Thanh asks Chuong if he has asked his employer whether they have a factory in California. But his question includes no overt reference to the recipient of Chuong's question (i.e., whom did he ask if there was a factory in California?) nor to the factory. With these references left implicit, the question in line 1 asks "But did you ask if California has?"

(17) (Vietnamese)

01	Thanh:	*Mà mày có hỏi bên Cali này có không?*
		but you have ask side/by California this have no
		"But did you ask if there is [the factory] here, in California?"
02	Chuong:	*Có không là cái gì ông nội?*
		have no is what grandpa
		"'There is' what, grandpa?"
03	Thanh:	*Có có chỗ mày mày làm giống dậy không á chứ?*
		have place you work similar that no
		"If there is a same place as where you worked."

If Thanh had included an explicit reference here, for instance the term *chỗ* ("place"), which he uses in line 3, it would have occurred between *có* ("have") and *không* (negative/question word) in line 1. Chuong responds in line 2 by repeating this framing, followed by *là cái gì* "is what?" In this way, he makes reference to an implicit or zero reference in Thanh's talk.

These examples illustrate a form of metalinguistic reference—reference to a referential act. The reference employs a technique of repetitional framing to draw the recipient's attention to a bit of immediately prior talk and to thereby treat it as insufficient.[8]

5.4.4 Redoing as a Tool of Metareferential Reference

A second practice of metalinguistic reference involves redoing the initial reference after a response to it has revealed a problematic understanding. These are cases of *third-position repair* (Schegloff 1992; see chapter 4, above). In example (18), a press relations officer in a civil defense headquarters (Annie) is asking the chief engineer (Zebrach) for information to be distributed to the media as to closures:

(18) Schegloff (1992, 1303)

01	Annie:	Which ones are closed, and which ones are open?
02	Zebrach:	Most of 'em. This, this, this, this (pointing).
03	Annie:	I don't mean on the shelters, I mean on the roads.
04	Zebrach:	Oh! Closed, those are the ones you wanna know about.
05	Annie:	Mm-hm.

Line 2 reveals to Annie that Zebrach has understood Annie's indexical expression "ones" (in line 1) to refer to the shelters, when in fact Annie meant to refer to the roads. Annie repairs the problem in line 3 and the course of action underway is then reengaged on the basis of the new understanding that Annie's correction provides for.

The materials we know from our own research contain many examples of such familiar practices. Consider example 19, from a telephone conversation between two Vietnamese friends.

(19) (Vietnamese)

01 Diep: *Vậy Lan đóng bao nhiêu tiền*

so Lan close how much money

"So, how much did you pay for that?"

02 Lan: *Không có đóng tiền gì hết á*

don't have to give any money at all

"I didn't pay anything at all."

03 Diep: *Không phải. Hồi đó Lan, Lan thi bằng lái á*

not that. that time Lan, Lan test license drive

"No. The time that you, you took your driving license test."

04 Lan: *Thi bằng hả?*

test license eh?

"Driver's test eh?"

05 *Thì mình đóng trong DMV bao nhiêu, mấy chục gì á.*

I give to DMV how much? Some tens something.

"How much did I pay at the DMV, forty something?"

Here when Diep asks her friend Lan how much she had to pay when getting her driver's license, Lan apparently misunderstands the reference and in line 2 responds by saying that she did not have to pay anything. Diep then corrects the reference by redoing it, this time making it explicit with *thi bằng lái* ("driving license test").

5.5 Conclusion

In this chapter we have shown how the referential capacities of language serve as both instruments and objects of accountability. As an instrument of accountability, language allows users to pick out and identify not only aspects of the phenomenal world but also of human conduct itself. This alone marks a watershed in the emergence of linguistically enhanced intersubjectivity as it allows speakers to direct people's attention to what someone has just done (or has done some time ago, or will do in the future; see chapter 7). But the corollary of this is where the real revolution happens. For it is not just conduct in general, but conduct pursued through the use of language that becomes the object of reference. The most basic practice here is the deceptively simple (but in the animal world apparently unique) use of what we call a metalinguistic index, which merely points to the just-previous talk. The prototype example of this is the interjection *Huh?* With this, a participant in interaction can effectively suspend the business at hand and attempt to redirect the other's attention not to what is coming next but, rather, to what has come before. This is a form of repair initiation. In our account it is equally important to see that it is an instrument of accountability. It is also a prime candidate mechanism among the origins of language (see chapter 8). From there we turned to increasingly complex ways in which the referential capacity of language is turned back upon itself. The most sophisticated of these involve kinds of *metareferential* reference—that is, practices by which speakers refer to prior acts of reference.

With the exception of interjections like *Huh?* making reference by means of language involves a formulation of the object of reference. In deictic reference, using words such as *this* or *that*, the formulation or characterization is minimal—for instance, merely picking out a region or an enumerable entity, or characterizing something as proximal or distal, animate or inanimate, and so on. When reference is accomplished with lexical nouns the characterization is more substantial. In many languages, such forms rarely occur in isolation (e.g., "Dog!") but are instead embedded in an encompassing noun phrase that is deictically selective, for instance, "that dog," "this dog," "my dog," and so on (Agha 2007). Other languages may require the use of numeral classifiers (Enfield 2021; Sidnell 2021). Notice that such lexical nouns necessarily involve some semantic characterization of the referent. The referent must be described one way or another, for example, as a "dog," "pet," "animal," "nonhuman" or "stray."

This means that when we use language, we are not merely identifying or drawing attention to referents, we are also thematizing and characterizing them. This introduces another crucial dimension to human intersubjectivity and, we will argue, to accountability. As with language more generally, the capacity for semantic characterization is both an instrument and an object of accountability.

6 Semantics for Accountable Description

> Besides the "immanent grammar" in every language, there is also in reality (i.e., even if not written) a "normative" grammar (or more than one). This is made up of the reciprocal monitoring, reciprocal teaching and reciprocal "censorship" expressed in such questions as "What did you mean to say?", "What do you mean?", "Make yourself clearer," etc., and in mimicry and teasing. This whole complex of actions and reactions come together to create a grammatical conformism, to establish "norms" or judgements of correctness or incorrectness.
>
> —Antonio Gramsci ([1929–1935] 2000, 354)

A sign can only be said to have meaning if it elicits interpretants, or meaningful reactions, which "point to" that meaning. We can, for example, say that the English word *fruit* means what it does because we see people react as if that's what it means. Suppose you phone a friend to ask if you should bring anything to the picnic. Your friend says *Bring some fruit.* You respond by bringing ripe apples, bananas, oranges, and pears. This response provides evidence that allows your friend to evaluate whether you understand the word *fruit* in the same way they do.[1]

The constant opportunity to test word meanings in this way exists because words occur in the enchronic frame, in utterances that are embedded in sequences of interaction. This frame for the social usage and circulation of language entails that words always occur as parts of actions that are both responding to something prior and initiating something next. The enchronic frame also entails that we are—again, always—socially accountable for our choice of words when we construct moves in interaction (Enfield 2013, chap. 3). When we say things in conversation, people's responses, linguistic and otherwise, publicly display their understandings of what we have said. Part of this displayed understanding will be an evaluation of the appropriateness of what we have said. As in any socially cooperative activity, with each move we are accountable for the appropriateness of our contribution. This accountability is a defining aspect of norms.

We submit that norm-grounded accountability in social interaction is ultimately what regiments the conventions of linguistic meaning. This is done proximally through the always-present possibility of holding people to account for the appropriateness of their usage of a word or phrase, and distally through the aggregate effects of this possibility in the historical conventionalization of form-meaning mappings in language communities (see Enfield 2014, 2015).

Because linguistic signs as form-meaning mappings are both revealed and constituted by the responses they elicit—in the enchronic frame of social interaction—and because we are accountable for social actions done by the use of any sign in interaction, word meanings will necessarily be regimented by social norms of a general kind. Linguistic norms are social norms. More specifically, the semantic norms that underlie symbolic meanings, as in the definitions of words, are ultimately distilled from (i.e., caused by) norms of social interaction of a general kind (an argument similar to that made by Silverstein 1976; Slotta 2021).[2]

The evidence for this comes from cases in which people dispute the appropriateness of a word that has been used, effectively invoking linguistic norms of proper application to reference.

6.1 Norms and Word Meaning

The meanings of words are regimented by norms in the same ways as things like the proper handling of answering questions and other more obviously pragmatic practices. If you use a word, you are socially accountable for intending to convey its conventional meaning. So, if a mother describes her teenage daughter as "a child," she is accountable for having characterized her with the set of properties that this description conventionally implies.[3] As a competent user of the language, she can anticipate (subprehend) the sorts of inferences that people will make from her linguistic choices.

Consider the following example. Virginia, a teenage girl, is challenging her mother's decision not to allow her to work in the family shop. Virginia cites the fact that Mom had allowed her sister Beth to work in the shop at a similar age. In justifying her decision, Mom obliquely describes Virginia as a "child" (line 4)—at line 5, Virginia objects to this.

(1)

01	Virginia:	Well Beth—didn't Beth get to work
02		before she was sixteen?
03	Mom:	No! I'd—I would let her wrap presents and packages
04		at Christmas and—times we needed somebody.
05		But people just don't want—children—waiting on them.
06	Virginia:	I'm not a child!

Virginia objects to the use of the word "child" and treats this as an incorrect description, thus explicitly holding Mom to account for having inaptly applied the word. The conventional meaning of the word provides a normative ground for disputing the appropriateness of Mom's utterance.[4]

Different words are often applied to the same referent, conveying very different construals: climate change *denier* versus *skeptic*, *freedom fighter* versus *terrorist*, *spared me* something versus *denied me* something, *dog* versus *mutt* (Enfield 2022). We select different linguistic formulations to fit the actions we mean to make with our utterances (Stivers 2007).

For example, our choice of words can portray an action as either intentional (culpable) or not (an accident). In an example we have discussed elsewhere, two children are playing with blocks. When the structure falls, one child screams, leading the supervising adult to complain, "Guys, too loud." This occasions an excuse from the child, who assigns responsibility to the other by saying "She poked it." The other child counters with "I tapped it." Here, the coded semantics of the English words *poke* versus *tap*—clearly different in meaning, but both plausible for reference to a class of situations—are used strategically for potentially consequential ends (Sidnell and Barnes 2013). Whether it is explicitly articulated as in this example (or not, as is more often the case), the linguistic expressions by which reference is accomplished in any instance of interaction are necessarily selected from among possible, contextually adequate, extensionally accurate expressions. For example, in the utterance "There's my dog," the word *dog* is selected from among equally possible, not-inaccurate expressions such as *poodle, pooch, pet, baby, boy, little guy, best friend, constant companion*, and so on (see Brown

1958a and b; Schegloff 1972, 1988a, 2000a; Stivers 2007; Sidnell and Barnes 2013).

Norms governing word selection operate over relations of granularity (*this week* vs. *today* vs. *this morning* vs. *at 9:30 a.m.*; *in the US* vs. *in Alabama* vs. *in Birmingham*), hyponymy/hypernymy (*dog* vs. *poodle*; *give* vs. *donate*), perspective (*my wife's nephew* vs. *my mother-in-law's grandson*; *chase* vs. *flee*; *buy* vs. *sell*), and others. In each case, the context-situated choice between equally accurate forms may be regimented by norms of appropriateness, which ultimately feed back into building community norms around what those words should be taken to mean.

What are word meanings? One way of answering this is to say that whenever a word is used, that word will dependably invoke a definable core idea in the minds of people who hear the word being used (Enfield 2015; 2022; Wierzbicka 1996; Goddard 1998). Someone who uses the word will be accountable for having intended to convey at least this core meaning, which is to say that the speaker will be unable to plausibly deny that they had wanted to invoke this understanding. Correspondingly, a listener will be accountable for having understood the word with that meaning, as long as they claim, explicitly or implicitly, to have understood the utterance in which the word was used (though there are complications; see below).

Suppose that Kim says to Pat "I'll bring some fruit," and Pat replies "OK." If in fact Kim arrives with muffins (or pumpkins), she will be accountable for having lied or misspoken (or could give the reason that she'd changed her mind or forgotten, appealing to some alternative normalizing of the problem). She won't be able to defend what she had said earlier, a fact that is explained with reference to a core meaning of the English word *fruit*. She can't, under normal circumstances, like Humpty Dumpty, say of muffins that *she* calls them a kind of fruit.[5] Something similar applies to Pat. If Kim brings a pineapple, Pat can later assert that he hadn't expected this (Kim hadn't said it would be a pineapple). But he can't later assert that he hadn't expected, for example, that Kim would bring something that people could eat (even though Kim hadn't specified it would be *edible* fruit). More accurately, if Pat were to assert that he hadn't expected it to be fruit, he would be accountable for the inconsistency between this and his explicit signal of understanding in the response he had given to Kim (he said "OK").

This is similar to an example given by Harvey Sacks: if I invite you for "drinks," you can expect no dinner, but if I invite you for dinner you can guess there will be wine and other drinks (and you can indeed express surprise, or sanction me, if I don't give you anything to drink). Drew (1992) provides examples from courtroom interaction. In one case, a question is directed to a plaintiff in a rape case: "Did he sit with you?" A polar question such as this invites a response with confirmation or nonconfirmation (e.g., "yes" or "no," confirming repetition; see Raymond 2003; Enfield and Sidnell 2017a). But instead, the plaintiff responds with "He sat at our table." At issue is the difference between the meaning of *with us* (implying he was part of our social unit) versus *at our table* (implying spatial proximity only, and indeed implicating, by choosing not to say "with us," that the person was *not* part of the social group at the table). In the example, the plaintiff is effectively refusing to be held to account for having said that *He sat with us*. This connects back to the above discussion of answering questions: if you don't answer, then inferences may be drawn, so in order to direct or contain those inferences you answer but modify.[6]

There are other occasions in which such discrepancies arise. Consider, for instance, the following case in which Jerry, who is at work, is talking on the phone to his wife Linda.

(2)

01	Linda:	Try and get home at a decent hour 'cause—
02	Jerry:	Yeah, be home by nine.
03	Linda:	No, get home pretty early, okay?

Linda uses the conventional expression "at a decent hour" in line 1. This conveys the idea of a time of day but does not specify what time that is. Jerry's response in line 2 provides such a specification. Jerry supplies an interpretation—or at least a claim—as to what the phrase "a decent hour" could mean. Linda contradicts this immediately with "no" in line 3, then adds "pretty early," another nonspecific descriptive time expression.

And there are occasions in which the normative appropriateness of a word's usage is bluntly questioned. For example, a grandmother calls out to her grandchild, "Careful! Don't run!" and the grandchild responds with "I'm not running, I'm jogging!" This case is about imposing a construal on a scene. The grandchild is effectively saying that she will not let *run* stand as a description of a scene that she might have some accountability for (e.g., if she trips and hurts herself, the grandmother might want to say that she should not have been *running*).[7]

Clearly there are important issues of entitlement here, matters of who has the right to say whether a description has been appropriately used or not. Such rights flow, primarily, from epistemic authority over the content of assertions. People are normatively entitled to describe their own private experience in whatever way they see fit (Sidnell and Barnes 2013). They may be called out for being dramatic or excessively humble, but they will not, under normal circumstances, have such first-person exclusive-access descriptions replaced or corrected by others. Rather, we find others merely highlighting or pointing to possibly inadequate wordings and leaving it to the speaker to correct for themselves. For example, in the following case from a radio interview, Hunter Harrison, a railway executive who served as the CEO of the Canadian National Railway and the Canadian Pacific Railway, among others, is describing how he got into the business. This he concludes by saying that he "developed some knack for it."

(3)

Rogers: So, when did you fall in love with it?
Harrison: Early on. Once I learned that, uh, oil and bearings
wasn't the thing I wanted to do. As I started
started looking around and seeing other opportunities,
all of a sudden fell in love with the business. I'm not
a buff or a foamer as I call them, but I have enjoyed
this business and I developed some knack for it.
And, uh, I've stayed with it for a long time.
Rogers: Some knack for it. I think some would say that's an
understatement.
Harrison: Well, thank you.

Rogers in line 8 thematizes the expression "some knack for it," and by suggesting that the expression does not apply—specifically, in the sense that it is "an understatement"—she conveys that Harrison has more than a knack for it, which Harrison in turn takes as a compliment (line 9). This case shows a typical pattern in which speakers tend to downplay self-praise, and in turn hearers may push back against such self-deprecation (Pomerantz 1978).

In striking contrast, speakers routinely engage in outright and overt correction of descriptions that pertain directly to themselves and their inner states. Consider the following case, in which the issue is the psychological state and behavior of the interviewee, actor Charlie Sheen.

(4)

01	Canning:	Your anger and your hate, I think, is coming off as
02		erratic to people.
03	Sheen:	Passion. My passion. It's all passion.
04	Canning:	Okay, your passion.
05	Sheen:	Yes.
06	Canning:	is coming off as erratic.

One issue here is whether the words *anger*, *hate*, and *passion* are applicable to what is being discussed (Sheen's state of mind and behavior). Another issue is who, of the two people in conversation, has the right to adjudicate that applicability, given that one of them has special access. It's Sheen we're talking about, not the interviewer. This reflects a general rule whereby people claim primary rights to say what they are thinking and feeling, what their motivations are, and so on (and notice that the interviewer immediately accepts the correction, incorporating it into her redoing of the question in the next turn).

A revealing exception to this rule is psychotherapy. In this context, the therapist is entitled and, indeed, expected to know the mind of the client or patient better than they know it themself (Peräkylä 2012). This unusual state of affairs arises in settings where a strong asymmetry of expertise or authority overrides a strong asymmetry of access.[8] Thus we find that therapists routinely treat the patient's description of their own experience as potentially problematic and requiring immediate correction (see Sidnell and Barnes 2013; Rae 2008).

Our larger point is this. It is not just that word meanings are regulated by norms but that people are motivated to enforce such norms even in the most mundane and informal of settings, and they are differently entitled to do so.

And it is not just others who hold speakers accountable to norms that govern the appropriate use of words. Speakers hold *themselves* accountable. This fact underlies a speaker's own selection of words in such a way as to ensure they are generally appropriate. The process is seldom visible to us, taking place in people's heads, but occasionally a speaker makes their self-accountability to such norms public in their own talk (often for sociopragmatic effects; see Jefferson 1974).

In example 5 below, two people are discussing the lecturers they have in university classes:

(5)

01 Bee: Yeah, this fella I have—"fella"—this man—
02 he has—who I have for linguistics is really too much.

Bee initially uses the form *fella* to refer to the instructor/professor of her linguistics class. She subsequently replaces *fella* with *man*. However, before she does this, she repeats the word *fella* as if quoting herself disparagingly, thereby commenting on her earlier usage and drawing attention to her own lapse and apparent violation of a norm: a person who is your professor is not appropriately referred to by a term such as *fella*. This is because the meaning of the word *fella* conventionally implies a level of informality and familiarity that does not properly characterize the relation between teacher and student. In this example, Bee calls attention not only to the norms governing the appropriate use of words but also, perhaps, to her own distance from the university environment and her lack of familiarity with its norms of social relations (see also Jefferson 1974, 1987).

In another example, Emma is talking to her sister Lottie about a man she admires. She first describes him as a "viral man" but, recognizing her error apparently as she says it, corrects to "virile," explicitly announcing that that was the word she wanted. In this way, Emma exposes her own error as well as the selection process behind it, one that seems to have led her to use a word that is not totally within the vernacular.[9]

(6)

01 Emma: Here's Mr. Black, older than any of them.
02 he's such a viral man—
03 virile—the—the word I wanted and, uh,
04 Barbara said he looked awfully bad.

To summarize, we have explored the fact that there are norms that govern word usage and meaning, and there are also norms governing the way in which such normatively prescribed word meanings are enforced and regulated. People not only hold others accountable to observing these norms, but also hold themselves so accountable. In these ways, the norm-governed flow of sequences of social interaction provides a matrix in which norms of language are used to regiment the use of language itself, and thereby concretize these norms in the form of semantic conventions.

Consider now the more complex case shown below. Here Betty is arranging to visit Ann, after not having seen her once-close friend for many years. The two are planning to get together in the early afternoon, after Betty visits with another old friend.

(7)

01	Ann:	Well, I'd like to see you very much.
02	Betty:	Yes. Uh—
03	Ann:	I really would. We could have a bite, and talk.
04	Betty:	Yeah. We—no! No, don't prepare anything.
05	Ann:	And uh—I'm not going to prepare,
06		we'll just whatever it'll be, we'll—
07	Betty:	No! I don't mean that. I mean—because, uh, she and I'll
08		probably, uh, be spending the day together, so, uh, we'll
09		go out to lunch or something like that.
10		So I mean if you, uh, have a cup of coffee or something,
11		I mean that, uh, that'll be fine. But, uh—
12	Ann:	Fine.

Here, as Schegloff (1987, 205) notes, "the issue appears to be the proper understanding of the term *prepare*." Ann's statement in line 5 shows that she understands Betty's injunction ("don't prepare anything") as meaning don't prepare something *special*. From line 7, Betty moves to correct this understanding, indicating that she meant don't prepare *any food at all*, as she will be eating with another friend.

Notice, then, that the speaker of the utterance that gives rise to a misinterpretation—"don't prepare anything"—is the one who clarifies its meaning, despite the fact that this has not been challenged or questioned by the recipient. Rather, the need for clarification comes to light because the recipient's response has embodied or displayed an understanding that the initial speaker takes to be mistaken or inadequate.

6.2 Norms and Explicit Metasemantic Discourse

When we think of metasemantic discourse, we might naturally think of the massive elaboration of prescriptive and descriptive work by educators and scholars in dictionaries, grammar books, and teaching materials in major languages such as Spanish, Hindi, Japanese, English, or Russian. But this kind of culturally elaborated discourse does not illustrate the foundational nature of metasemantic discourse of the kind we are focusing on here. Most languages exist perfectly well without dictionaries, and indeed without any system of writing. But they will all exploit the capacity for human language to serve as the means of its own explicit description in order for people to question or challenge the normative appropriateness of a word or phrase in situ. In the last section we reviewed examples in which such an orientation to the application of language in context is subtle and involves only a fleeting suspension of the business underway, without a detour into significant metasemantic discourse about the meaning or appropriate uses of words. But we also see cases in everyday interaction that take a step up from this, where interactants will make it their business to focus explicitly on the question of what words mean.

This kind of metasemantic discourse may arise spontaneously when the intended meaning of a word is not obvious. Consider the use of *done* in line 1 of the following conversation between Mike and his superior at an advertising agency, Stef:

(8)

01 Stef: So, when are you done with Porter?
02 Mike: What do you mean by "done"?
03 Stef: Like when can you remove yourself from that process?

In this case, rather than respond to the question Stef has posed, Mike asks for clarification as to the intended meaning of "done." The problem here appears to result from the somewhat unusual grammatical frame in which the otherwise perfectly ordinary word "done" is embedded. Mike is reporting to his superior at work, Stef, about his progress for the month. Stef's unidiomatic "When are you done with Porter" (which is essentially equivalent in meaning to "When *will you be* done with Porter") suggests some kind of special meaning for the word "done," one with which Mike is unfamiliar. And notice that the repair produced in line 3 provides additional support for such an analysis: Stef does not say "like when will you be finished," but rather "like when can you remove yourself from that process," thereby again employing "account-manager talk," that is, talk designed to give the impression of a specialized, technical register.

Or consider the following case in which a lawyer is questioning a plaintiff in a sexual assault case and picks out the word "hysterical" for elaboration:

(9)

01 Plaintiff: I couldn't tell if she was crying or not, I—
02 she was a little hysterical, uh, they're real weird
03 and then I just—by the time I left I never saw
04 her crying.
05 Lawyer: Well, let me see, hysterical? What do you mean by that?
06 She's shaking?

And there are many more mundane cases like this. These appear frequently, for instance, in interactions among young children, presumably as a result of the fact that word learning is shaped by the particularities of the child's home environment (e.g., whether the child's parent is an astronomer or an engineer, etc.). In each of these three examples from conversation between five-year-old children, one participant momentarily turns the business of the interaction to defining a word that has just been used (*cab*, *tartar sauce*, and *barnacle*, respectively):

(10)

01 Hillary: I went in a cab so I didn't have to drive my car.
02 Meg: What's a cab?
03 Hillary: It's um—it's a kind of, ah, thing that takes some people somewhere they want to go.

(11)

01 Sam: Tartar sauce.
02 Jane: Tartar sauce. What is tartar sauce?
03 Sam: It's, um, a sauce th—the sauce you put on French fries.

(12)

01 Shane: Vickie, you barnacle head.
02 Vicky: What's a barnacle?
03 Shane: A barnacle is, uh—a thing that attach to ships.

It's not just children who engage in these sorts of exchanges. The following case is from an interaction between a mother and her teenage daughter. After the daughter (Virginia) uses the word *gwaff*, the mother initiates repair (with "what?"). As this elicits only a repeat of the word rather than an explanation of what it means, Prudence (Virginia's sister-in-law) then explicitly asks for a definition:[10]

(13)

01	Virginia:	I know, but all the rest of my—people my age are gwaffs.
02		I promise, they are sick.
03	Mother:	They're what?
04	Virginia:	Gwaffs.
05	Prudence:	What's a gwaff?
06	Virginia:	Gwaff is just somebody who's really . . .
07		I just—s—immature.
08		You don't want to hang around people like that.

Word teaching often involves a similar kind of metalinguistic discourse. Consider an example from child-caregiver interaction:

(14)

A one-year-old child is looking at a book, pointing to the page with a picture of a kangaroo.

The parent says "It's called a kangaroo. Kangaroo." The child says "Roo." (Clark 1999, 6)

In (14), the child does not ask for the word, but the adult nevertheless volunteers it, and so turns the exchange into a language teaching moment. "Now, with the new word, the child can recognise, and draw attention to, a new distinction. Often such moments involve more than just learning a new label. Adults will say things about a new word's relationship to words that the child already knows" (Clark 1999, 6). Consider, for instance, the following, in which the adult attempts to clarify the relationship between two words or ideas:

(15)

A three-year-old child points to a picture of an aardvark and says "That's a kangaroo." The parent says "Well, that looks like a kangaroo but it's called an aardvark." The child says:

"Aardvark." (Clark 1999, 7)

Metasemantic inquiries are often, as in the cases so far discussed, directed toward individual lexical items. They may also be directed to larger stretches of talk. This takes many forms. Here we briefly describe one practice. Consider the examples given as 16–18, each of which features a deceptively simple mechanism for calibrating understanding of language in the moment-by-moment flow of interaction. This mechanism—marked by an arrow in examples (16–18)—is accomplished using a reflexive metalinguistic idiom ("you mean") to which is added a formulation of the upshot of what a prior speaker has just said. This functions as a "candidate understanding," which is offered up for the first speaker's confirmation or disconfirmation.

(16)

01	Emma:		Well—I don't know, Lottie, what's gonna happen.
02			I don't know.
03	Lottie:		Yeah, well.
04	Emma:		I'm not so scared as I was. But I'm still kinda scared.
05	Lottie:	→	You—You mean to stay there alone?
06	Emma:		No.
07	Lottie:		Huh?
08	Emma:		Oh no, not to stay alone. There's so much love around
09			this place it's u—unbelievable

(17)

01	Madie:		I was just wondering you know . . . could—
02			do you think you might . . . want to rent . . .
03			you know like the bottom part of your . . .
04			garage like to me for a while, or something
05			like that.
06	Marsha:	→	. . . Well. . . . Oh—you mean for living in, Madeline?
07	Madie:		Yeah.
08	Marsha:		. . . It's just . . .
09	Madie:		Not possible, huh?

(18)

01	Croft:	He's quite a character. I like him. I get along with him,
02		but his wife scares me to death.
03	EJ:	Mm, well—what's the ma—what she do to you?
04	Croft:	Well, she complains that I didn't know that the padlock

didn't work on the front door, not supposed to walk
through that. I mean, just little items like that she—
EJ: → You mean they padlock their front door, they—
Croft: Oh yes, the whole front area. Well, they own three houses
and three lots.

In each of these otherwise diverse cases, someone is telling something to a recipient and the recipient responds with a question of the form "You mean" + [inference]. Thus:

You mean + [to stay there alone?]

You mean + [for living in?]

You mean + [they padlock their front door?]

In the first case, Emma is telling her sister Lottie about her plans for Thanksgiving in the wake of a recent break from her husband. When Emma uses the adjective *scared*, she does not specify what she is scared of, or about. With the "you mean"-prefaced response, Lottie offers a possible formulation of this based on a range of available inferences (i.e., that Emma, having separated from her husband, is now staying at their summer home by herself). This gives Emma the opportunity to confirm that this is an accurate and adequate formulation of her meaning, which in this case it is not: in lines 8–9 she disconfirms that interpretation. This simple clarification of understanding could not have been achieved without the reflexivity of language.

In the second case, Madeline has called her friend's mother to ask if she can rent the garage. The tortuously produced request in lines 1–4 is marked by many perturbations and delays. Moreover, when Marsha does not immediately respond (a delay that typically serves as a harbinger to rejection), the request is extended first with "like to me," then with "for a while," and finally with "or something like that" (see Sidnell 2010, 157–158 for analysis). When she does respond, Marsha does not reject the request outright but asks for clarification by offering up a candidate understanding for Madeline's confirmation.

In the final case, Croft is telling EJ about the couple with whom he is lodging. Croft explains that the wife complains about him, mentioning a specific occasion in which he appears to have attempted entry via a padlocked door. EJ asks "You mean they padlock their front door?" offering an inference that Croft is then able to confirm. Notice that the question headed by "you mean" elicits an explanation, showing that this practice can index something counterintuitive, or perhaps normatively aberrant.

These examples illustrate what we call the *ampliative* character of human shared understanding powered by the reflexive capacity of language. In each case, the "you mean" question, by making an inference explicit, adds something to what has been said and offers it up for confirmation or disconfirmation. By raising inferentially derived information for confirmation, the addition to the domain of shared understanding is obvious: it is now in the participants' common ground that Emma is not scared to stay on her own, that Madeline is asking to live in Marsha's garage, and that Croft's hosts secure their premises with a padlock. This kind of calibration and grounding happens continually in interaction. Such practices make inferences and understandings explicit and necessitate the shaping of these understandings into propositions.

6.3 Three Types of Metasemantic Discourse

The kinds of exchange just reviewed may seem trivial. But they do not differ in kind from forms of discourse to which we attach great importance. Indeed, they show that those forms of discourse are rooted in totally ordinary forms of everyday talk. Metasemantic discourse is ubiquitous. Here we note three broad types.

First, word meanings rise to the level of public debate for a variety of reasons. Often, these discussions are a part of larger social and political movements (see Enfield 2022, chap. 8). For instance, for at least fifty years there has been intensive discussion about the meaning of English words relating to gender and occupation such as *actor* and *actress*, *waiter* and *waitress*, *policeman* and *policewoman*. There are questions about the appropriateness of using *man* to refer to "humankind" and of using *he* or *his* to refer to a group that includes both men and women. And most recently, the use of pronouns has attracted public discussion, for example in relation to the use of *they* to refer to a single person as a way to avoid classifying them by sex. These sorts of discussions show how norms relating to social, gender, ethnic, or racial group identity are subject to reflexive reconsideration and change in language.

A second set of issues concerns expert knowledge and authority. In many domains, the meaning of a word is decided by experts who can lay claim to exclusive, expert knowledge (Putnam 1975). Some of the examples Putnam discusses—such as *gold* and *water*—are controversial, given that these are everyday words that English speakers learn to use in everyday ways (Wierzbicka 1985). But other cases are clearer. For instance, it is not for laypeople to decide what *schizophrenic* or *bipolar* mean. They may use these words freely in their own talk but a psychiatrist who overhears them would be within her rights to tell them that they'd used these words incorrectly. Obviously there's a gray area here, with cases in which the authority of experts erodes and popular usage rules (e.g., *oedipal* or *narcissist*), or where authoritative sources such as the *Diagnostic and Statistical Manual of Mental Disorders* can significantly change their definitions of terms for disorders from decade to decade partly, it would seem, in response to lay usage (see Haslam 2016).

A third set of issues has to do with legal discourse. Much of law hangs on what the people involved determine words and expressions to mean (Waismann 1945; Hart 1961; Dworkin 1986; Hacking 1986, inter alia). This means both the specialists in their interpretation of legal texts and the "reasonable persons" whose interpretations set the coordinates for accountability. In an example of a legal quarrel over the interpretation of a word, in 2017 the US Senate Intelligence Committee interviewed former FBI Director James Comey about a conversation with Donald Trump, then US president. The conversation was in relation to Lieutenant General Michael Flynn, who had left the White House after misleading the vice president about his relations to foreign governments, including Russia. Trump said to Comey—whose bureau was investigating possible foreign interference leading to Trump's election—"I hope you can see your way clear to letting this go, to letting Flynn go." The key issue for the Senate was to determine whether Trump's saying this should be understood as an order (or a threat) to Comey. That determination hung on the meaning of the word *hope*.

For instance, during his questioning of Comey, Senator James Risch quoted from Comey's written statement:

> I want to drill right down as my time is limited, to the most recent dust-up regarding allegations that the president of the United States obstructed justice. And boy you nailed this down on page five, paragraph three. You put this in quotes. Words matter. You wrote down the words so we can all have the words in front of us now. There's twenty-eight words there that are in quotes and it says, quote, "I hope," this is the president speaking, "I hope you can see your way clear to letting this go. To letting Flynn go. He is a good guy. I hope you can let this go." Now those are his exact words, is that correct? (*The Spokesman-Review*)

After Comey confirmed that this was correct, Risch continued with, "He did not direct you to let it go." To this Comey assented, saying, "Not in his words, no." And Risch followed with, "He did not order you to let it go," to which Comey replied, "Again, those words are not an order." (*The Spokesman-Review*)

In this first exchange, then, Risch focuses on the question of what Trump can be reasonably said to have done (e.g., "direct you," "order you") in producing the utterance under discussion. Risch's questions and Comey's responses to them suggest that for these participants this was to be decided by considering the words that were produced. In this view, an utterance has the effect it does, and performs the action that it does, solely by virtue of the words of which it is composed.

This example well illustrates both the limits and potential importance of naïve native metapragmatic discourse. In trying to decide what a given utterance was doing, people often make recourse to the "literal," surface meaning of that utterance, going by its constituent parts. This introduces obvious distortions. One distortion derives from the fact that such naïve analysis typically treats words as the sole or most important conveyor of meaning and, moreover, as simply naming their referents rather than making a choice to frame them in one way and not others. Consider Risch's subsequent questioning of Comey:

> No. He said, "I hope". Now like me you probably did hundreds of cases maybe thousands of cases charging people with criminal offenses. And of course, you have knowledge of the thousands of cases out there that—where people have been charged. Do you know of any case where a person has been charged for obstruction of justice or for that matter any other criminal offense where this—they said or thought, they hoped for an outcome? (*The Spokesman-Review*)

What has happened here? Risch has identified the verb of the matrix clause (*hope*) as somehow the key to understanding the utterance as a whole. And this verb is, it seems, taken to point to some inner state. Specifically, it is treated as referring to a real and discrete mental state of *hoping*. Following through on this logic, Risch frames the question as being about a person's culpability for *hoping* for a certain outcome.

Clearly there's a muddle here. Part of the problem lies in treating the word *hope* as a label for a preexisting and independent mental state.[11] Part of the problem involves failing to recognize that the meaning of an utterance is the product not only of lexical semantics but of grammatical relations. *Hope* is a complement-taking predicate that belongs in a form class with other predicates such as *wish* and *fear* as well as *think* and *believe*. Like these other predicates, *hope* can take a finite complement. Like *wish* and *fear*, it is implied that the subject of *hope* does not know whether that which is expressed by the complement is or is not the case (or will be so in the future). Unlike *wish*, *hope* can be used with complements that express future possibilities. (Thus, "I hope he will be there" but not "*I wish he will be there."[12]) The point is obvious: the meaning of a word like *hope* is not to be identified with its lexical semantics alone, still less with that to which it supposedly refers (even if it could be decided what that might be!). The meaning of *hope* must also derive from its grammatical properties. Bloomfield suggested that a grammatical analysis along these lines could potentially provide an exhaustive account of a word's meaning (Bloomfield 1933, also [1943] 1970). For our argument, it's sufficient to note that important aspects of meaning are conveyed not by lexical semantics or referential extension, narrowly conceived, but by grammatical relations. These are largely beyond the level of awareness of users of the language.

6.4 Conclusion

We have considered semantics, broadly construed. The semantic properties of language are harnessed as instruments of accountability. They provide the means to formulate and thematize people's conduct, to place that conduct "under a description," as Anscombe put it (Anscombe 1957, 28 and passim; 1979; see also Sidnell 2017). At the same time, semantic aspects of language are themselves an object of accountability. Just what a word or phrase or larger stretch of talk "meant" is something that people frequently discuss. We have considered the apparent limits of such naïve analysis, and we finished by suggesting that users of a language rarely appreciate the meaningful grammatical properties of words or of language in general. We turn now to the contribution that grammar makes to our understanding of intersubjectivity.

7 Grammar for Intersubjectivity

Let us take stock. Our investigation of the conditions and consequences of language began in chapter 1 with the concept of intersubjectivity. Then, in part I, we covered elements of the infrastructure for human interaction, without which language could not operate as we know it. These elements are *status* (chapter 2), *participation* (chapter 3), and *enchrony* (chapter 4). Then in the first two chapters of part II, we linked these elements of the infrastructure for language use to two core structural-functional features of language: *reference* (chapter 5) and *semantics* (chapter 6). This sets the stage for a discussion of *grammar*. This is the system perhaps most widely regarded as "core" to language, at least by twentieth-century linguists and cognitive scientists, yet perhaps least obviously linked to—as either consequence of, or condition for—intersubjectivity in human social affairs.

We outline the status of grammar in relation to our two core questions in this book. First, in what way does primary intersubjectivity make language possible? Second, in what way does language then enhance human intersubjectivity? We shall see that grammar presupposes intersubjectivity and in turn offers highly developed, fine-grained, and powerful means for enhancing human intersubjectivity and transforming our social lives.

7.1 Grammar Contains Diverse Subsystems with Diverse Natures and Functions

"Grammar" refers to a heterogenous set of rules and signs that people use productively in organizing words, phrases, utterances, and discourses.[1] It is more of an ecosystem—a set of interrelated systems—than a single, coherent system. The lexicon, which we discussed in the last chapter, consists of *open-class* stand-alone signs such as the nouns and verbs of English. They are called open-class because they belong to classes that are large and open to receiving new additions, for example when new words are coined, or borrowed from other languages. In many languages, there are thousands of nouns, with new ones being adopted every week, while the set of pronouns numbers only a handful, and may see changes such as additions, mergers, or subtractions only once in a century or two.

Grammar can be said to contain everything *other than* the open-class vocabulary items of a language.[2] This covers at least two main types of functional structure. First are the generative patterns and constructions embodied in phrase structure rules, idiomatic templates, and other types of ordering principles that enable people to generate novel utterances using their inventory of words. These principles allow us, for example, to distinguish between *Kim saw you* and *You saw Kim*, or to describe an animal as not just a horse but more specifically as *that horse*, *my horse*, or *the black horse*. Second is the system of closed-class grammatical markers, for example, pronoun and demonstrative paradigms, case-marking affixes, agreement markers on verbs, and other kinds of bound morphological marking (e.g., evidentiality and modality), to be discussed below.

In the next three sections, we explicate three central functions of grammar: grammar links reference and predication (clause structure); grammar links narrated events to speech events (displacement); and grammar links language to language (reflexivity).

7.2 Grammar Links Reference and Predication: Clause Structure and Beyond

A simple but powerful function of grammar is to bring together the two major functional elements of language introduced in the last two chapters. Grammar supplies rules for taking words and combining them into propositions. These propositions are the basis of *the clause*, which has subject-predicate or theme-rheme structure.[3] A clause refers to something and then characterizes it in some way. Take the clause *Kim is asleep*. First, it points to *Kim*, setting this referent up as a theme or subject of joint attention (see chapter 5). Next, it introduces some new piece of information, something *about* Kim. This is the predicate. To understand this sentence, you focus your attention on Kim and then you learn that she is asleep.

Every language provides resources for organizing words into clauses in this way (though the specific rules for doing this differ across languages, sometimes radically; Sapir 1921; Bloomfield 1933; Greenberg [1966] 2005; Croft 2001; Evans and Levinson 2009; Dixon 2010; *inter alia*). In terms of our interests in this book, the simple function of linking the act of referring to the act of predicating provides an essential and powerful building block for the construction of intersubjectivity. Using the productive mechanisms of the clause, a person can not only draw someone's attention to something, they can convey new information about that thing. This increases shared information, which is a prerequisite for intersubjectivity.

But shared information itself is not sufficient. As we explained in chapter 1, for information to become intersubjective, the information must be shared *and* it must go on the shared record *that* the information has been shared (Lewis 1969; Clark et al. 1983; Clark 1996; Stalnaker 2002). We might both know that Kim is asleep, but it is only when the sharedness of the information is in the open—for example, when I say to you that she's asleep and you nod in understanding—that both of us become mutually accountable for our subsequent actions in relation to the proposition that Kim is asleep.[4] For example, when the fact is in the common ground, we don't have to explain why we are whispering or tiptoeing around the house. Conversely, we won't be surprised to be sanctioned for playing loud music.

So, grammar carries a huge functional load in building intersubjectivity by increasing common ground through productive generative mechanisms for linking subjects to predicates in novel propositions.

The power of syntactic structure in the clause is ramped up enormously when we add another level of capacity for specifying further details in our descriptions of events and situations. The subject of a clause isn't just a noun but a noun *phrase*. A phrase can be simple or complex. So instead of *Kim is asleep* we could say *My niece is asleep* or *My poor tired niece is asleep*. Because grammar is semantic in nature, the expressive alternatives provided by grammar have the same implications for intersubjectivity as the kinds of lexical contrasts we introduced in the last chapter. Grammar radically amplifies those possibilities for selecting among alternative framings of the situations we want to describe.

In the last chapter, we discussed the implications of word selection for accountability, for example of the participants in a situation being described.

Recall the children disputing what happened just before a toy block structure tumbled down: A: *She poked it!* B: *I tapped it!* The generative nature of grammatical constructions means that a speaker's possibilities for alternative framings of a situation are not limited to the existing inventory of relevant words the language happens to furnish (e.g., *poke* versus *tap*), but are literally infinite.[5] This is important, in part, because accountability attaches not to brute facts but to formulations of those facts (consider, "You didn't talk to that man" versus "You didn't say hello to your friend").

Grammars also provide the means for extending basic clause structure with increasingly specific information in associated structures such as prepositional phrases or their equivalents. For example, we could specify the location of the event: *Kim is asleep* <u>*in the next room*</u>. The discipline of linguistics is heavily concerned with describing and understanding grammatical mechanisms for constructing clauses and their constituent phrases. Much of that research focuses on the organization of referential information in propositions that describe, or make claims about, events and situations beyond the specific context in which the proposition is being made.

We will not review the formal complexities of grammar's potential for infinite expression here (see, e.g., Van Valin and LaPolla 1997; Dixon 2010; *inter alia*). Our interest is to explicate the sense in which that potential plays an important role in enhancing human intersubjectivity.

7.3 Grammar Links En to Es: Displacement and Beyond

The capacity of grammar to link reference and predication is just the first of three important intersubjectivity-enhancing aspects of grammar. The second component is the role that grammar plays in linking the *narrated event* to the *speech event*.

Following Jakobson (1971), we adopt the following terminology. Given a linguistic proposition, the *narrated event* (En) is whatever event or situation the proposition refers to or is about. In this technical terminology, "event" covers any state of affairs, not just dynamic events. The *speech event* (Es) is the situation of speaking itself. If I say *Kim was asleep*, the past-tense form of the verb *was* is an explicit signal that the time of the narrated event (Kim being asleep) is *before* the time of the speech event (now). We use the term *origo* (Karl Bühler's term; Bühler [1934]1982) to refer to a point in time-space-personnel (*here-now-me*), such that the Es origo (time of speaking, place of speaking, participants in speech act) may be different from the En origo.

The very possibility of communicating about events and situations beyond the here and now is one of the uniquely defining features of human language. Hockett referred to this as *displacement* in his 1960 article on the design features of language. Animal calls lack this capacity for displacement.[6] For example, a vervet monkey can make a call that means something like "Threat from the sky!" (the so-called eagle call; see Cheney and Seyfarth 1992). The call signals that there is a threat *here, for me,* and *now*: there is no distinction between the narrated event and the speech event. The vervet cannot communicate to its associates that, say, there was a threat from the sky earlier that day or that the threat is somewhere else than here (or, further, that there was *no* threat from the sky yesterday or that the caller thinks there is a threat but isn't sure). In systems of animal communication, the event of communicating (Es) has the same coordinates in terms of space-time-personnel as En, the event being communicated about. In other words, nondisplacing systems of animal communication do not have means for distinguishing Es from En.

By contrast, human language allows us to make propositions about events and situations that have different space-time-personnel coordinates from the speech event. This feature of language enables us to inform people about things that happened beyond their experience but which it might be useful to know. Suppose someone tells you *There is a wasp nest in that hedge*. Those words will prompt you to take care to walk around the hedge, not barge through it, and thereby avoid harm. Or suppose you collect some *Colocasia gigantea,* a wild forest tuber, not knowing exactly what it is, and you are told *My uncle nearly died from eating this*. Those words will relieve you of learning that lesson the hard way. This powerful capacity for informing others is often foregrounded when people emphasize the special utility of language. People can tell you things so you don't have to learn them from experience. This is central to the cumulative nature of human culture.

But saying things to people does more than simply update them on the informational content of propositions. It is part of the relationally strategic building of common ground, which is a form of investment in a social-affiliational economy. Gossip and related forms of talk are oriented to the sharing of stances and the management of reputation. We not only report things that might have happened, we also use lexical and grammatical resources to frame and portray those events as good or bad, right or wrong. Telling people what others have said or done is an opportunity to share social values, to praise or to blame others, and to manipulate social affiliations by declaring and securing allegiances (Gluckman 1963; Haviland 1977; Goodwin 1991; Dunbar 1993; Enfield 2013, 2022). Similarly, the combination of infinite expressivity and displacement in language is the foundation of narrative more broadly (most gossip is already narrative, of course), which in turn is at the heart of social sense-making. Grammar supplies many of the tools we use for these functions.

Several of the subsystems of grammatical marking found in the world's languages are dedicated to marking relations between Es and En.[7] For example:

Tense: *Kim is asleep* versus *Kim was asleep*

Aspect:[8] *Kim fell asleep* versus *Kim has fallen asleep*

Modality:[9] *Kim is asleep* versus *Kim might be asleep* or *Kim must be asleep*

Evidentiality: *Kim is asleep* versus *Kim is apparently asleep*

Another such grammatical category is egophoricity (San Roque, Floyd, and Norcliffe 2018). This is a grammatical system that marks canonical versus noncanonical alignment between the person of a clausal subject (first versus second) and the speech act function of the clause in its context (assertion versus question). Consider table 7.1.

In English, the person of the subject and the speech act function of the utterance are independently expressed, but they are not independent in practice. Two of the four possibilities in the table are pragmatically marked, meaning that they are far less frequent in usage, and when used they can

Table 7.1
Logical combinations of subject (first versus second person) and speech act functions (assertion versus question) in a sample of clauses with the predicate "hungry" in English (San Roque et al. 2018, 66)

	First-person subject	Second-person subject
Assertion	I am hungry.	You are hungry.
Question	Am I hungry?	Are you hungry?

sound unusual and thus invite special inferences. These are the shaded cells, in the bottom left and top right of table 7.1. Because a person has privileged access to knowledge and experience about themselves, this leads to two natural correlations: (1) Propositions with first-person subjects will usually occur in the speech act function of assertions, and (2) propositions with second-person subjects will usually occur in the speech act function of questions. In most languages, evidence for these correlations is found solely in frequencies: instances of the shaded cells will be far less frequent than the unshaded cells. See table 7.2.

As table 7.2 shows, of a sample of "hungry" clauses in English with first-person subject, 100 percent of these are assertions and none are questions. And of those with second-person subject, questions (i.e., "Are you hungry?") account for 95 percent of cases. Put another way (and recall we are focusing only on clauses with first- or second-person subjects), if an assertion is made that someone is hungry, only seven percent of these will have a second-person subject, while 93 percent will be first person, and if a question is asked as to whether someone is hungry, then 100 percent of these will have a second-person subject.

This pattern is surely not surprising, but we note that in English it is evidenced only in these frequency measures and is not marked in the grammar. But in some languages, there is overt grammatical marking of this asymmetry (San Roque et al. 2018). With a system of grammatical marking of *egophoricity*, speakers will use one type of marking to signal that the correlation of person and speech act function is the canonical one (so, marker 1 is used with both first-person-subject statements and second-person-subject questions) while the other type of marking is used in the noncanonical, unexpected situation (i.e., marker 2 is used with first-person-subject *questions* and second-person-subject *assertions*).

Table 7.2

Distribution of subjects (first versus second person) and speech act functions (assertion versus question) in a sample of clauses with the predicate "hungry" in English fiction texts (San Roque et al. 2018, 66).

	First-person subject	Second-person subject
Assertion	171	12
Question	0	249

Systems such as these add information to the basic subject-predicate content of clauses and sentences, giving further weight to the intersubjectivity-enhancing effects of language. By adding more specific information, the accountability implied by the speech act is fleshed out. In languages with systems of evidential marking, speakers are accountable for marking the evidentiary basis for what they're saying: if they haven't seen the event for themselves, they should make this explicit using a "hearsay" or "inferential" marker or something similar. In languages with multiple distinctions in tense (e.g., recent past versus past), speakers are accountable for marking this distinction accurately. At the same time, such marking will also allow people to highlight or downplay accountability for narrated-event participants.

To summarize, grammar gives us the capacity to productively build novel phrases, clauses, and sentences, and to do so in ways that (1) orient to the distinction between Es and En and (2) finesse that distinction in subtle ways. These functions of language massively amplify the language's capacity to enhance human intersubjectivity. Grammatical systems such as tense-marking, evidentiality-marking, egophoricity-marking, and modality-marking are dedicated to specifying certain kinds of links between a speech event and a narrated event. These semantic specifications are conveyed by grammatical elements in closed-class systems. They convey information that is not limited to En, the narrated event.[10] They provide information about Es, the speech event, and therefore they orient to language itself. This takes us to a third relation between grammar and intersubjectivity, the language-unique function of *reflexivity* (Hockett 1960).

7.4 Grammar Links Language to Language: Reflexivity and Beyond

So far, we have talked about structure at the level of clauses and sentences. But a focus on this level alone—endemic in formalist linguistics—would be insufficient. This is because people don't talk in isolated sentences (Foley and Van Valin 1984). Instead, in the enchronic context of language usage, language comes in the form of grammar-hewn chunks in sequences of social interaction. If you utter a clause or a sentence, it will never simply point to and characterize some state of affairs. The utterance will always be occasioned by something in a social interaction, and it will always occasion something next. The speech event is not an isolated point. It is a link in a temporal chain. Every contribution is at least two things beyond its informational content: (1) Looking back, it is a response to, and interpretation of, what just came before, and (2), looking forward, it sets the conditions of relevance for, and interpretation of, what comes next. This is most obviously true when each move is produced by two different people, as in a back-and-forth conversational structure like a question-answer sequence. But it is also true when the series of moves is produced by a single person, as when a speaker uses language to do something that takes more than one turn, such as giving a set of instructions or narrating a series of events.

So, our first point is that in discourse, people don't just understand each contribution but understand that it orients to its "cotext" (i.e., linguistic environment, as opposed to "context" more generally). Language is always linked to language.[11]

The idea that language can be about language is made most explicit in the phenomenon of quoted speech. Suppose it turns out that Kim wasn't asleep after all and when I find this out, I sanction you for misleading me. I say, *You said that Kim was asleep*. My utterance refers to your previous utterance. This kind of function is not available to even the most sophisticated animal communication systems. No vervet monkey can hold another vervet monkey to account for, say, giving a wrong signal or not giving it quickly enough. No bee can communicate to another that their waggle dance was unclear or false. But with language, humans do this sort of thing all the time.

To talk about the things that others say, we need a dedicated set of syntactic strategies for *embedding* others' utterances into our own utterances. In the example just given, the verb *say* takes the complementizer *that*, which in turn takes as its complement a finite clause whose propositional content is that of your original utterance. A good deal of the machinery of complementation strategies in the grammars of the world's languages is dedicated to supplying ways of saying things about things people say (Vološinov [1929] 1973; Goffman 1974; Lucy 1993; Dixon and Aikhenvald 2006; Rumsey 2021; Zuckerman 2021).

Beyond the grammatical tools for quoting others' speech, there is a much more expansive and open-ended set of mechanisms for linking language to language. These are the mechanisms for building narrative.

Narrative is key to human collective sense-making and it would be impossible without the grammatical mechanisms that languages provide for achieving cohesion over larger texts (Bruner 1991; Ochs and Capps 2009; Dancygier 2012). As we have noted, people typically produce series of utterances rather than isolated signals. We are about to see some of the ways in which the resources of grammar provide explicit ways to connect moves in linguistic sequences.

First, we note that the primary, highest-level kind of glue that joins these utterances together is not marked in grammar but is imposed by the human capacity for assuming associations among adjacent signs, and for using general processes of inference to arrive at the most likely interpretations. Suppose we simply juxtapose two propositions (uttered at midday):

Kim is asleep. She worked the night shift last night.

Human inferential processes are such that we will not take these propositions to be unconnected. We will infer, by abduction, what the relevance of the second part is to the first part, and thus what the speaker must have wanted to say by adding it. Most likely, we would take the statement that Kim worked the night shift last night to be an explanation of, or justification for, *why* she is sleeping at midday. Note that this kind of inference draws heavily on broader, intersubjectively-shared common ground (e.g., knowledge about shift work, about daily sleep needs and normative sleep patterns, etc.) to figure out what kinds of connections between propositions are most likely to be intended and why.[12]

Our always-on abductive inference engines—using a sort of association heuristic—will go a long way to providing a general degree of coherence among the many propositions that together, and in a given order, make up a discourse. But the resources of grammar can add a great deal of further detail and refinement, giving speakers the means to constrain hearers' interpretations. In the example we just gave, we included one such grammatical resource that lends structural cohesion between the two clauses. That is the resource of pronouns. Look again at the example:

Kim is asleep. She worked the night shift last night.

In the second clause, the word *she* is understood to refer to Kim, though in itself it only specifies a third-person feminine singular. It picks up its reference precisely by its placement in a discourse sequence (Halliday and Hasan 1976). This word is one of a closed class of grammatical words in the set of personal pronouns in English.

While this use of pronouns is a universal strategy for textual cohesion, the semantic distinctions made in pronoun systems can vary widely, depending on the (often historical) cultural concerns of the communities in which the language has evolved. The English pronoun system is relatively simple among languages in the number and type of semantic distinctions made (singular/plural; first, second, or third person; and masculine/feminine in third person singular). Different languages will add different additional information.

Some languages add distinctions relating to kinship. For example, in some Australian Aboriginal languages there are alternative dual pronouns, one pronoun marking a pair of people who are in a generationally harmonic relationship (e.g., in the same generation such as two brothers, or two generations apart such as a grandmother and granddaughter), the other marking that they are in a disharmonic relationship (e.g., adjacent generation such as a father and son) (see, inter alia, Sapir 1913; Hale 1966; Merlan and Heath 1982; McGregor 1996; Evans 2003a, 2003b). This is information about participants in En, the narrated event, but the pronouns also function to link utterances that stand in a certain relationship to each other within the enchronic course of Es (Silverstein 1987; see also Nichols 1984).

Another kind of meaning distinction that language might mark in pronouns relates to politeness. In many European languages, there is a choice between distant/formal and familiar/informal pronouns in second person reference (e.g., *tu/vous* in French, *du/Sie* in German; see Brown and Gilman 1960). This kind of distinction is oriented more to Es, the speech event, signaling something about the relationship between the speaker and addressee, independent of what is being said about the narrated event.

There are many other kinds of grammatical device for explicitly marking cohesion across clauses. One way is to explicitly mark the relationship of conjunction between clauses or sentences. In the following examples from English, the underlined words mark distinct meaning relations among propositions: *She got on her horse <u>and</u> rode home; He was <u>so</u> upset <u>that</u> he had to be alone; His crop failed <u>because</u> the soil was poor; She forgot the words <u>so</u> she hummed the tune.* In another kind of case, a relationship between clauses is marked by a combination of a word and an affix: *He was eat<u>ing</u> lunch <u>when</u> they arrived.* All languages have these kinds of interclausal linkers. Some have patterns of marking that operate at greater distance across many propositions in a narrative sequence. Many languages of Papua New Guinea feature a grammatical pattern known as clause chaining.

In a typical Papuan clause-chaining construction, a series of events is predicated in a number of verbs used in sequence. All of the verbs are marked with a grammatical suffix, which marks that the verb is *medial* in the chain (i.e., non-final). Medial verb marking can specify agreement-related values of person and number of the narrated-event participants but does not specify tense, aspect, or mood. For their tense/aspect/mood values, these medial verbs depend on the inflectional marking of the final verb in the chain. Here is an example from Ku Waru:[13]

(1) (Ku Waru) (Rumsey, Reed, and Merlan 2020, 4)

Olyo med maket-ma-nga pu-p

we down.there market-plural-genitive go-medial

Kalyip baim te-p

peanut buy do-medial

No-b pilawa lyi-p no-

eat-medial flour.balls get-medial eat-medial

Pu-mulayl

go-future/first-person-plural/definite

"We'll go down to the markets and buy peanuts and eat them, and get some flour balls and eat them and then we'll go."

All languages have ways to indicate such interclausal relations, lending structure to narratives and increasing informational specificity. This increases common ground among interactants and also adds specific content to the accountability being intersubjectively established, both among speech event participants and among narrated-event participants.

So far, we have concentrated on grammatical means for achieving cohesion in series of propositions in narratives, where each move represents one of a number of happenings in the narrated event. This is textual cohesion. It is needed if we are to make sense of monologues. But in the enchronic context that dominates our use of language, cohesion is needed not just between propositions in a narrative, but between propositions in back-and-forth sequences of interaction, that is, in the speech event involving two or more participants.

The most fundamental expression of cohesion between propositions produced by different participants in a speech event comes under the rubric of *sequence organization* (Schegloff 1968, 2007a; see chapter 4, above). The assumption of relevance allows us to infer meaningful relationships between propositions in a sequence produced by a single person. Similarly, in dialogue, the assumption of relevance allows us to infer the probable meaning relationship between adjacent propositions. Consider the following:

1. Person A: *Is it cold there?*
2. Person B: *It's freezing.*

Again, these aren't disconnected propositions. We understand the second utterance, from Person B, to be not just an assertion about some narrated event but an *answer* to the question posed by Person A. Now notice that in the enchronic context of dialogue, the connection of relevance between the two utterances isn't only a cue for our interpretation of what the second proposition means in relation to the first. It also generates real-time rights and duties for the speech event participants. Once a person addresses a question to somebody, as in the example just given, then the addressee of that question is subject to an interpersonal imperative that Schegloff (1968) termed *conditional relevance* (see chapter 4). Because of this imperative, Person B in our example is momentarily obligated to address the question. They should answer it if they can. If they can't answer it then they should at least provide a response (even if that is "I don't know"). If they do not—for example by simply not responding, or by saying something entirely unrelated to the question—then the answer that line 1 had projected will be "officially absent," which in turn will mean that Person B may be held accountable for not having fulfilled the duties implied by their status as "addressee of a question" (see chapters 2 and 3). The complexities of conditional relevance go beyond our current scope. We raise the issue here to point out that conditional relevance is often brought about, regulated, or refined by the grammatical resources of the language being used.

7.5 Grammar Matters for the Human Cognitive Capacity for Intersubjectivity

Research on language and social intelligence over the last few decades has discovered multiple kinds of relationship between language and the psychological basis for intersubjectivity, often referred to as theory of mind. In their 2014 review of this field of research, de Villiers and de Villiers (2014, 313) define theory of mind as the "ability to understand that other people have minds, and those minds contain beliefs, knowledge, desires, and emotions that may be different from [one's own]." This is obviously an important part of the human infrastructure for intersubjectivity. Research on the link between language and theory of mind has found evidence for three main types of connection, focusing on the development of theory of mind in children (much of which takes place in the first four years of life).

A first link between language and theory of mind concerns the semantic content of frequently used words and expressions about other people's inner states or first-person experiences. We use words with children early on to discuss what they want and think, how they feel, and so on. Such talk draws our attention to others' inner states and also encourages us to build hypotheses—largely based on our own experience—as to what those states are and what they are like, and ultimately to build folk-psychological models of the inner life of others (see Nelson 2005; Hutto 2008).

A second link is in the information about other minds that typical conversational language use provides, assuming the kinds of relevance-based, associational inferences described above. De Villiers and de Villiers (2014, 314) cite the following exchange:

1. Person A: *I am going to have chocolate spread on my toast.*
2. Person B: *That's Marmite!*

Marmite is a savory food spread made from yeast extract. It is a salty, yeasty concoction that might, to the untrained eye, look like chocolate spread. The exchange provides evidence that Person A's utterance in line 1 is not merely a proposition but is treated as evidence of the person's belief, namely, that the spread is made of chocolate. Here, Person B discerns that it is a false belief, and they act accordingly, informing Person A of the truth. (This action itself reveals an attention and orientation to others' inner desires; namely, the assumption that people would want their false beliefs to be corrected.) This is grounded in the fundamental idea that speech acts have conditions of appropriateness, many of which have to do with the speaker's state of mind (Austin 1962; Searle 1969). Thus, when a person makes a statement, asserting a proposition, this conveys—by presupposition—that they believe the statement to be true. (Of course, they may be lying, but our vulnerability to being fooled by lies is grounded precisely in our usual assumption that people's statements are grounded in their sincere beliefs.) Listeners learn to adopt the working presuppositions as a matter of principle in the cooperative context of conversation (Grice 1975; see also Harris 2005).

A third link between language and theory of mind brings us back to this chapter's focus on grammar. This link is made by the grammatical resources that a language provides for expressions that explicitly represent our own and others' states of mind. These features of grammar are akin to, and often identical to, the structures used for quoting the speech of others, mentioned above. An example is *I forgot that Kim was asleep*, where the proposition that Kim was asleep is embedded in the clause as a complement of the verb meaning "to forget."

Languages provide means for *embedding* one clause in another.[14] This function is especially important for intersubjectivity because it is our tool for thematizing and characterizing people's beliefs and desires, and thereby for publicly coordinating around the things that people think, want, and feel. These are the grammatical resources for *complementation* (Dixon 2010, vol. 2, 370ff).

Verbs of cognition such as *want* and *know* may occur in simple clauses, each taking a noun or noun phrase as their complement:

I want <u>water</u>.

You know <u>the answer</u>.

But they also often take a complement that is a clause in itself:

I want <u>to leave now</u>.

You know <u>that Kim is asleep</u>.

And such embeddings can be further embedded:

You know <u>that I want</u> to leave now.

The ability to control this kind of complex embedding of others' inner states and perspectives presupposes the ability to conceptually isolate and manipulate others' inner states and perspectives. In the example just given, the speaker linguistically represents not only what the main subject (you) knows (a first-order representation of another person's inner state) but also the subject's representation of another person's inner state (a second-order representation: you know what I want). But crucially for the argument we are making in this book, not only do these kinds of linguistic structures *presuppose* certain capacities for intersubjectivity, they also help to *enable* the underlying human capacities for intersubjectivity. For young children who are developing those capacities, "mastery of the appropriate linguistic structures gives the child a new ability to reason about the contents of others' minds" (de Villiers and de Villiers 2014, 314). But taking this out of the domain of individual psychology and into the public realm of enchronic interaction, we can say that such structures massively amplify the capacity of people to build intersubjectively shared knowledge and to hold one another accountable. This aligns with our argument that, firstly, primary intersubjectivity makes language about other minds possible, and secondly, in turn, that as language about other minds develops, this further enhances human intersubjectivity, taking it to a new level.

While the complex technical details of syntactic embedding are beyond our scope here (see Foley and Van Valin 1984; Gívon 1984, 1991; Van Valin and La Polla 1997; Dixon 2010) it is nevertheless worth pausing briefly to consider the ways in which the subtleties of meanings in different mental predicates and speech act verbs are associated with differences in grammatical structure and behavior. Consider the English verbs *want* and *wish*. Their meanings are similar enough that they are often interchangeable without significant consequence:

John wants to excel.

John wishes to excel.

Other contexts, however, reveal an important difference:

I wish that I had met my grandmother.

**I want that I had met my grandmother.*

As Dixon (2014, 32) writes, "The verb *want* is only used of something which is achievable, and it is restricted to a complement clause introduced by *to*, as in *I want to meet Barack Obama*. There is a more wistful sense associated with *wish*, relating to something that is not possible, and it will then take a complement clause introduced by *that*" (see also Dixon 2005). In addition, while *wish* subsumes "want," it also contains additional semantic components relating to *knowing* (Wierzbicka 1988, 132ff), and so it is compatible with English *that*-complements in a way that *want* isn't.

We can gain further insight into this point by looking at inner-state verbs that can occur with both types of complement structure. Consider the verb *remember*:

I remembered to lock the door.

I remembered that I had locked the door.

(Cf. *I wanted to lock the door* vs. **I wanted that I had locked the door.*)

The event of remembering is relevant both to achievable subsequent action (and hence compatible with a *to*-marked complement) and to the situation of a proposition being the object of my thought or knowledge (marked with *that*).[15] Other words, like *hope*, introduce further nuance: although one can say *John hopes to excel*, one cannot say **I hope that I could have met my grandmother*. This is because *hope* cannot take complements that express counterfactual propositions.

The point is this. Meaning differences between the inner-state predicates that create, maintain, and regulate our shared conception of mental life are not solely in the semantics of the relevant words but also in the grammatical frames within which the words may occur. Such inner-state predicates no more point to preexisting mental states than speech act verbs point to preexisting types of act (see Enfield and Sidnell 2017a and 2017b). Rather, such words and grammatical constructions analyze, construe, and construct mental states and, in turn, structure the ways in which we talk and think about how the mind works.[16]

7.6 Intersubjectivity Is Not for Minds but for Agents

The significant scholarly literature about language and theory of mind has focused on questions of individual psychology, albeit psychology in the social domain. Here is de Villiers and de Villiers's (2014, 325) summary conclusion of their review of the role of language in theory of mind development:

> A breakthrough arises with the child's ability to represent, via complex grammatical language, the contrast between the content of his own beliefs and knowledge and those of others and to use this contrast to reason about others' behavior. Being able to talk about minds leads to a richer theory, one that continues to help make sense of social situations.

This summary is in line with our enhanced intersubjectivity argument, but it puts more emphasis on the development of individual mental capacities. It is about the individual's ability to "represent," "reason about," "talk about," and "make sense of" others' beliefs and desires. But there is something important beyond these individual capacities and we want to draw attention to it. That thing is action. Having linguistic means for representing the thoughts, desires, and words of others is not only a way to understand other people, it is a way to make claims about their social accountability, and in turn to position them, and oneself, in a set of fundamentally social relations. And then one acts accordingly.

So, if someone says *You know that Kim is asleep*, they are not just stating what you know, they are drawing your attention to something that is relevant for evaluating your behavior. For example, if you are not taking care to be quiet when moving about the house, the statement *You know that Kim is asleep* might be an admonition. This might contribute to cementing the speaker's publicly registered stance that you are an inconsiderate person, which in turn may affect your relationship with that person and, more broadly, your reputation.

And if someone says *She thought it was chocolate spread*, they are not just stating what she thought. They might be explaining or justifying her behavior to somebody else. They might be using this as evidence for her outsider status (if, say, she is an American lodging with a family in England). They might be using this to make a case that she is careless or dim-witted. These aren't just matters of our individual capacity to mentalize about others' minds. Social intelligence is more than this. It includes the motivations and capacities for creating and manipulating statuses and relationships, and for enabling social action.

Through processes of the kinds we have reviewed, a child who is acquiring language and other aspects of social competence will draw on language, and to a large degree on grammar, to achieve the enhanced intersubjectivity that characterizes human social life. In so doing, as Katherine Nelson (2005, 28) puts it, the child is thus able to "enter the community of minds." We agree, but we would add that it is not enough to be a mind among minds. The child who has developed linguistically enhanced intersubjectivity is enmeshed in frameworks not only for mutual knowledge but for mutual values, ethics, and actions. The community they enter is a community of accountable agents.

7.7 Where Grammar Comes From

In this chapter we have been concerned with the role that grammar plays in enhancing human intersubjectivity. By focusing on features of grammar such as pronominal systems, markers of interclausal relations, speech act markings, and complementation strategies, among much else, we see how grammar achieves the following things, above and beyond what words alone can achieve:

- linking words to other elements in a clause
- linking speech events to narrated events
- linking clauses to clauses in larger sequences; forming texts such as narratives
- linking utterances to utterances (e.g., in reported speech)
- representing and expressing people's inner states, such as beliefs and desires
- regulating statuses of speech event participants

Our observations about the intersubjectivity-enhancing properties of grammar in this chapter have been limited to structures and processes in two distinct temporal-causal frames. One of these is the synchronic frame, to use de Saussure's term. Synchronic descriptions of the elements, structures, and rules of grammar as a system that hangs together are entirely abstract compared to the temporal dimensions of language use. We have described grammar in synchronic terms, but at the same time, for the purposes of understanding grammar's role in intersubjectivity, and in particular its relation to matters of the speech event, we have also found it necessary to invoke the enchronic frame, the frame that foregrounds the role of language in constructing moves in sequences of social interaction (see chapter 5). The enchronic frame is, we argue, at the confluence of linguistic and social processes occurring in all other causal-temporal frames. The enchronic frame should be privileged in any analysis of language and intersubjectivity because it constitutes the bottleneck through which all language (and all signs) must pass.

Let us put this into the broader context of research that seeks to explain why grammatical systems are the way they are. Attempts to explain observed language structures will begin with synchronic description, as we have just defined it. While we include the synchronic frame in a set of diverse causal-temporal frames for language (Enfield 2014, in press), it stands apart from other frames because it is the only frame that is in fact neither causal nor temporal. It is purely relational.

Once you know the set of relational patterns and rules that hold among linguistic signs in a language system, the question then arises: How to explain them? Different kinds of answer will appeal to different causal-temporal frames. One kind of answer appeals to the diachronic frame. This refers to the population-level historical processes by which grammars evolve, over decades and centuries. As an example, consider the robust finding from descriptive work on the world's languages that there is a strong correlation between (1) the relative order of verb and object in clauses and (2) the relative order of adposition and noun in adnominal phrases. English conforms to this correlation, with verb-object order in clauses (*Kim*$_{subject}$ [*saw*$_{verb}$ *you*$_{object}$]) and preposition-noun order in adnominal phrases (*at*$_{preposition}$ *home*$_{noun}$). Japanese shows the opposite correlation, with object-verb and noun-postposition order. Most languages (approximately 94 percent) follow one or the other of these patterns.[17] There are different explanations for why this is.

A diachronic explanation might begin by observing that in many languages adpositions historically evolve from verbs. What used to be a verb-object structure in a clause retains its original ordering but changes its structural/functional role to become an adposition-noun structure. What used to be a verb is now a preposition. In this account, the correlated structures are not independent. One has been derived from the other, over a timescale longer than an individual human lifetime.

Another kind of explanation appeals to processes or patterns in the *microgenetic* frame. This frame is the causal-temporal setting of cognitive processes: the processes of perception, attention, memory, motor control, and reasoning that individuals draw on in the real-time production and comprehension of language, or indeed of any waking moment of action or interpretation in a sociocultural setting. When Greenberg ([1966] 2005) first noted systematic correlations of word order patterns in languages of the world, he referred to the dominant correlated typological patterns (noted above) as *harmonic*. This idea was grounded in principles of individual cognitive processing of language. A harmonic relation in grammar is "connected with the psychological concept of generalization" (97).[18]

This idea is rooted in the general tendency for any entropy-resisting system such as a human body and brain to minimize energy cost for maximal benefit. In the first half of the twentieth century, the linguist George Zipf formalized and tested a "principle of least effort" in individual language processing, linking it to emergent properties of linguistic systems such as the robust tendency for more frequently used words to be shorter (see Zipf 1935, 1949).

Data such as the relative frequency of use of grammatical options in a language are seldom included in synchronic descriptions. For example, a synchronic grammar would state that in Sanskrit, nouns may be inflected for three numbers: singular, dual, and plural. It is implied that the three options are structurally equivalent. But the three options are used with radically different frequencies, introducing strong asymmetries in their values in the grammatical system. Greenberg ([1966] 2005) reported that 70 percent of Sanskrit noun tokens in texts are in the singular, 25 percent are marked for plural numbers, and only 5 percent are marked for dual numbers. This conforms with a cross-linguistic universal tendency for the singular to be the unmarked member of the set of markers of grammatical number. This often means that the singular is not only the most typical or most often encountered number, it is also often literally "unmarked" in that no formal marker is present (e.g., *dog* versus *dog-s* in English, where singular takes "zero" marking). This insight was made famous by Zipf more than seventy years ago, when he noticed that more frequently used words require less effort to process (Zipf 1949; see also Bybee 2010).

Hawkins (2004) includes [singular → plural → dual] among a set of *performance frequency rankings* that are accounted for by pressures on the microgenetic processing of language. Hawkins argues that patterns of grammar correlate with patterns of cognitive and motor performance. He proposes three ways in which the organization of grammars tends to maximize frequency: (1) minimize the domains to be marked (e.g., if there are three numbers, only mark two of them, and ensure that the most frequent one does not get any marking), (2) minimize the linguistic forms to be processed (i.e., make the markers short), and (3) "select and arrange the linguistic forms so as to provide the earliest possible access to as much of the ultimate syntactic and semantic representation as possible" (9).

Let us consider an example of how this kind of efficiency-based microgenetic argument has been applied in explaining grammatical structure. Since Greenberg's pioneering work comparing grammatical structure across the world's languages, it has been known that languages have a preference for prefixes over suffixes. Fifty-five percent of languages prefer suffixes, while only sixteen percent prefer prefixes (Martin and Culbertson 2020, citing Dryer 2013). It has been argued that this asymmetry follows from biases in human perception: "the beginnings of words are most salient to the human speech-perception system, and this privileged position is reserved for the most important content: the stem" (Martin and Culbertson 2020, 1107). This argument appeals to the microgenetic frame of language processing, in which words are processed in milliseconds, as they are incrementally perceived. When you are listening to a person speaking, as information is coming in, millisecond by millisecond, you need to identify each next word in a sequence as soon as you can if you are to understand what's being said. In this context, if the stem comes first in an inflected word (as in *walk-ing*), the uniquely identifying part of the word comes first. You hear *walk* before you hear *-ing*. This favors faster word recognition and processing, and it decreases costs for language users, likely resulting in fewer incorrect understandings.[19]

We want to emphasize that this account is not purely about individually-situated cognitive processing. It is also supported by arguments from the enchronic frame (in ways that are hardly recognized in the psycholinguistics literature). If a certain organization of grammar tends to make it more difficult for hearers to recognize words in the flow of time, this would not only require more effort for listeners to process speech, it would also make such processing more prone to error. Considered purely from a microgenetic point of view, this creates an efficiency problem. But considered in the enchronic context of the speech event, with its socially interacting participants, the problem is not only about efficiency. If it leads to more disfluency, this would be expected to impact the always-relevant social-relational aspects of the speech event as well.

Himmelmann (2014) invokes the enchronic frame in his argument that suffixation is preferred from the point of view of the listener whose task includes, among other things, projecting the course of a current speaker's turn for the purpose of participating in the turn-taking of free-flowing conversation (see chapter 4). He presents an argument for the historical development of suffixes that is grounded in the two-party interplay of turn-taking and its implications for the prosodic and grammatical dependency of markers and the elements they mark.[20]

In a similar vein, Roberts and Levinson (2017, 402) develop an account of the historical evolution of constituent order across languages in the context of processing pressures that are "imposed from turn taking in conversation." An agent-based modeling experiment provides support for their contention that "interactive turn-taking in conversation must impose constraints on cognition" and "these may have implications for the way in which languages change over time" (424). This view is in line with the argument long promoted by Schegloff (1989, 2006, inter alia), that the structures of grammar are shaped in the interactive context of conversation. Thus, Schegloff says (1989, 144), we must understand those structures to be "adaptations to the turn-at-talk in a conversational turn-taking system with its interactional contingencies."

We now close this brief review of some interacting biases that operate upon processes of language usage and transmission to give rise to grammatical structure as we know it. We conclude that ultimately, grammar is a cultural-evolutionary outcome of social interaction under intersubjectivity. The biases that have determined its shape are those set by the enchronic context in which intersubjectivity operates. Grammar is the population-level output of biased evolutionary processes that must pass through the enchronic bottleneck (chapter 4), and in turn grammar constitutes the shape of that bottleneck for subsequent iterations of the process.

7.8 Conclusion

We have argued that the foundations of grammar in human language are intertwined with human intersubjectivity, in the two senses that we have pursued through this book: (1) Grammatical systems would be impossible without primary intersubjectivity first being established in the social contexts in which language is used and evolves, and (2) once grammatical systems are established, they amplify, enrich, and enhance human intersubjectivity in turn. As de Villiers (2007, 1858) writes: "the interface between language and Theory of Mind is bidirectional." Because grammar is the outcome of linguistic usage in social interaction, any existing grammatical system has necessarily been created by interactions among people in a world of intersubjectivity. In turn, a grammar provides coordinates for, and constraints on, the localized nature of intersubjectivity and the horizons of interactional possibilities for social associates who inherit that system.

Postlude

8 Conditions and Consequences of Language

We have argued for two interconnected claims: first, a primary form of human intersubjectivity has made language possible in our species, and second, the emergence of language has, in turn, been transformative, resulting in the enhanced intersubjectivity that characterizes human societies and cultures today. Despite the impressive social capacities of other species, such as the other great apes, their interactions are not undergirded by the same elaborate interactional architecture of accountable intersubjectivity seen among humans.

The difference is a consequence of language. Our species' dynamic, language-enhanced form of intersubjectivity is foundational to all forms of historically constituted and transmitted knowledge. All macro-level institutions and networks, we argue, are ultimately the product of context-situated enchronic encounters among invested human agents.

There would be no enhanced intersubjectivity without accountability and there would be no accountability without language. Accountability and language are thoroughly intertwined, in at least two ways. One is that language use and linguistic meaning are themselves normatively regulated and the distinctive, indeed unique, properties of language make it subject to a starker and more fine-grained form of accountability than any other form of human conduct. This is partly because of the discrete, granular properties of utterances that allow for precise segmentation, isolation, and identification of what has been said, even to the scale of individual grammatical markers.[1] The other is that language can figure as an instrument, rather than object, of accountability, as we explored in part II.

Let us finish not with a summary of our argument but by highlighting some of its implications.

8.1 The Linguistic Foundations of Language?

There is a paradox at the heart of our argument. We have argued, in essence, for the *linguistic foundations of language*. But how can language be grounded in language? There are in fact two puzzles here. First, if language is a system whose elements derive their sense, in part at least, from contrast with other elements, how can the system exist in advance of the elements of which it is composed? Second, if a definitive property of language is the presence of metalanguage, how could there be language about language if there were not already language? We want to offer a solution to this paradox, one that solves both its constitutive puzzles: no elements without system and no language without metalanguage.

Concerning the second of these puzzles, many contemporary thinkers assume that "reflexive discourse is a superficial supplement to language itself, one which could be removed without seriously affecting language" (Taylor 2000, 486). Searle (1995, 73), for instance, suggests, that language is "self-identifying." As Taylor (2000, 487) puts it:

> [Searle] argues that institutional facts such as money, property, marriage, etc., require language—indeed "are constituted by" language. However, he claims that language is the exception to this. For, although also institutional, linguistic facts—such as the fact that *comely* is a word of English or that it means "pleasant to look at"—do not require language. Language does not require (meta)language in the same way that other institutional facts do. Instead, language is precisely designed to be a self-identifying category of institutional facts.

But Taylor asks, "How far could human language have evolved without the use of reflexive discourse?"

> Naturally, this question sounds somewhat paradoxical—as if the suggestion were that early humans could not have developed "primary" language (language itself) until they had first developed a higher order language, a metalanguage, for talking about primary language. Of course, one might then want to ask: "Well, if that's true, then how could they have developed metalanguage until they had first developed meta-metalanguage for talking about metalanguage?" The absurdity of the regress is obvious. (489)

Paradox or not, it is undeniable that a language in which one could not perform the elementary metalinguistic operation of agreeing with what someone else has said, or asking for clarification, or indeed asking what a particular gesture or word was intended to convey (let alone "mean"), would hardly be worthy of the name.

We think that a solution to the riddle can be found in the universal organization of *repair* (section 4.5), and in particular in the discoveries that (1) *Huh?* is a word and, (2), it is found in every known language (Dingemanse, et al. 2013). The universal distribution of this word is explained by convergent evolution. We speculate that *Huh?* is "as close as you can get to the core of the human faculty for language" (Enfield 2017, 207). Could *Huh?* have been the first word, and the fount of language?

Imagine the scenario with which we began this book: a time in which protohumans communicated using voluntary, purposive gestures like those of contemporary nonhuman primates such as bonobos and orangutans (Rossano 2013; Tomasello 2008). These gestures could be used to make requests, to initiate play, to get another's attention, to beckon, and to offer. These gestures were, as researchers have shown for other primates, used in sequences of social interaction. For instance, a beckoning gesture addressed to a particular other could establish the relevance of a specific form of response, for instance compliance by moving toward the beckoning individual. Where a relevant response was not produced, the beckoner might produce a second attempt or might move to a location that was more visible to the addressee and thereby attempt to secure their attention.

Such gestural communication would have provided the requisite conditions for the use of *Huh?* Imagine an early human is addressed by a gesture, but fails to understand its intended import and instead utters an interjection of puzzlement in the sequential slot where their behavioral response should have occurred. The interjection is metacommunicative insofar as it is *about* the communicative action just made by another agent. This is the wedge that introduces metalanguage into a not-yet-linguistic system. Then imagine that this response promotes the redoing of the gesture. In turn, the link between the interjection of puzzlement and the redoing of the not-yet-understood gesture becomes ritualized and so may be used again to achieve the same end—a repetition of the previous communicative move—either by this protohuman or by another who happened to witness the events unfold. We can imagine here a shift from learning via ontogenetic ritualization to imitation, that is, conventional transmission, the beating heart of language and culture.

Would this constitute language? No, but it would explain how you can have metalanguage prior to the establishment of language as we know it. This is because, as we explained in chapter 5, *Huh?* is a metalinguistic vocal point or deictic. It orients to an immediately prior communicative act.

This leads to the other of the two component puzzles: how system-like relations can arise from such item-based beginnings (see Enfield 2014 on the "item-system problem"). The key is to understand that once a slot is established for an expected action in a recurrent communicative sequence, alternatives to that action—as long as they occur in the same slot—derive meaning purely from the fact that they are alternatives to a known or expected point of reference (cf. Darwin 1872, 50ff on the principle of antithesis in the expression of emotion). In this way, once a protoword like *Huh?* was established, it created a slot into which other interjections might meaningfully occur. The possibility of distinct elements occurring in the same syntagmatic slot creates the potential for one of the fundamental features of a linguistic system: paradigmatic alternation (de Saussure 1916; Zipf 1949). We have known of such grammar in the structure of interaction (i.e., syntagmatic-paradigmatic organizations of interactional moves) at least since Schegloff described the case of "uh-huh" as an alternative to the initiation of repair (Schegloff 1982; Sidnell 2009b). So, we can imagine that once *Huh?* was established, an opposing form could easily occur in that slot, indicating the "antithesis" of *Huh?*—that is, conveying the meaning that the prior move has been understood and that compliance is to follow.[2]

This sort of simple paradigm of opposed, minimally semantic forms would constitute a basis for language. This basis is not in the logical concepts of reference and predication or in the individual cognitive capacity to represent events and situations. Instead, it emerges from universal structures of semiotic action in the enchronic frame of move-by-move accountable interaction, of the kind illustrated, for example, by the two men sawing timber in chapter 1. A simple paradigm of responsive interjections might thus serve as initial steps toward grammar (see Hurford 2012 for an account of the evolution of grammar beginning with interjections). Subject to laws of semiotic conventionalization, over time these interjections would take on more semantic nuance and content. And these forms would provide jumping-off points for expansions in the form of additional slots and paradigms, kicking off open-ended processes of grammaticalization that underpin the complex grammatical systems seen in the world's languages today.

8.2 Implications and Contributions of Our Argument

Before concluding, let us draw connections to some parallel lines of work where we feel that our account can contribute value.

8.2.1 The Intelligence for Language

What happened in our evolutionary past such that we possessed the cognitive prerequisites to make the steps to language that we have described? We have presented arguments and evidence that the human capacity for language is grounded in a rich social-interactional infrastructure, comprised of diverse elements.[3] Recall the first of our two main claims in this book: that language presupposes primary human intersubjectivity, which is inherent even in our most basic forms of joint activity, such as the two-person timber sawing discussed in chapter 1. This claim is consistent with an influential line of thinking in research on the evolution of primate intelligence, which has argued that what sets humans apart from our closest relatives—who, like every other species, *lack* language—is not our intelligence but our *social* intelligence.

Ethologists in the 1960s were trying to understand the origins of higher primates' intelligence. They realized that the evolutionary pressure for ever-better brains was not coming from any special new challenges presented by the physical world. Alison Jolly introduced the idea that the evolutionary context for increasing primate intelligence was not in the physical world but in "primate social life" (1966, 506). Similarly, neuropsychologist Nick Humphrey asked why humans are "much cleverer than they need be" and answered that "the higher intellectual faculties of primates have evolved as an adaptation to the complexities of social living" (1976, 316).[4] The ensuing *social intelligence hypothesis* (Byrne and Whiten 1978) explored these connections. When primatologist Robin Dunbar saw a correlation between increased primate brain size and increased social group size, he hypothesized that language emerged as part of a social-cognitive revolution in our species. Riding on increased psychological capacities for tracking and understanding multiple embedded social relationships and perspectives, language provided a way to scale up the individual's number of social relationships beyond the overly time-demanding limits set by physical grooming (Dunbar 1993, 2003). Work in this vein has reframed the function of language similarly to the way Jolly and Humphrey reframed the function of primate intelligence. The idea is that we will only make progress when we understand that the function of language/intelligence is primarily social, not primarily informational/logical.

So far so good. But the line of research on language opened up by Dunbar has not progressed, for two reasons. The first is methodological. This line of work has assumed that the social functions of human language can be studied in the same ways as the social functions of other forms of animal communication. It has been concerned with questions like what people talk about, who people talk to, and for how long. To take one example, Dunbar, Duncan, and Marriot (1997) studied human conversation using the methods that Dunbar was trained to use in his pioneering work on the social behavior of gelada baboons in the Ethiopian highlands (Dunbar 1985). The researchers sat near people in cafeterias, bars, and trains, taking field notes on their conversations through "direct auditory monitoring" rather than audio- or videorecording the interactions:

> To assess the content of conversations, the general topic of the speaker's utterances was scored as an instantaneous scan sample taken at 30-second intervals: the topic of conversation at the moment of the sample was judged by the immediately preceding period of conversation. (1997, 234)

Leaving aside the difficulties of determining the "topic of conversation" at any given moment, this method has no way of measuring or analyzing the workings of the enchronic infrastructure for human interaction that are central to the management of intersubjectivity, as we have shown in this book. The ethological approach to language has, perhaps paradoxically, never appreciated that human social interaction is conducted in radically different ways from that of other primates precisely because language as we know it is the medium for interaction.[5] Language has been treated as a scaled-up tool for performing just the same functions as physical grooming: managing and tracking social affiliation and reputation. The ethological approach has not looked at the far richer measures of language's social functions that have been explored over the last half century in sociology: turn-taking, sequence organization, repair, and enchronic accountability. In this book, we have placed those things front and center.

The second limitation of Dunbar's line of research on the evolutionary foundations of language is that it doesn't explain why the grammatical structures of human languages are as complex and diverse as they are (Hurford 1999, 2012). Michael Tomasello, another evolutionary psychologist, developed a proposal that addressed this by distinguishing between three levels of complexity in language structure and correlating these with three broad classes of social function that language can have (Tomasello 2008):

Simple grammar ↔ requesting

Complex grammar ↔ informing

Fancy grammar ↔ sharing

But this theoretical proposal is highly abstract and needs to be fleshed out conceptually and studied empirically. In this book we have made some steps in that direction. In chapter 7, we outlined some elements of what might constitute "fancy grammar" in more specific terms than those offered by Tomasello (2008), though we are still far from providing the needed empirical demonstration of Tomasello's conjectures.

In sum, while research grounded in ethology and the study of primate social behavior and intelligence has made extraordinary progress, we have here outlined paths that this work is yet to take. We believe that these paths will lead to considerable advances in our understanding of the foundations of human language through an account of its social-interactional underpinnings that adequately addresses the structural complexity of the phenomena.

8.2.2 The Social Logic of Meaning

The idea that language is a system for carrying out logical operations on information has been around as long as philosophy itself. But language in real life doesn't tend to conform with textbook principles of logic. Consider the following exchange:

Professor: Where is your assignment?

Student: I've been ill this week.

In purely logical terms, the student's response is a non sequitur, but in natural conversation it is perfectly apt. In 1975, the philosopher Paul Grice sought to reconcile the mismatch between the known principles of formal logic and the principles that people apply in understanding utterances in everyday conversation. Grice (1957) had earlier established that meaning in language involves not just the semantics of words but also a higher-order consideration of the intentions of the person speaking. His work took the question of the understanding of language—which was widely regarded as a matter of logical calculation by rational individual agents (in the En domain)—and reframed it as an operation by which interpreters must take into account the presumed intentions of speakers (in the Es domain), and by which, in turn, speakers must design their utterances assuming such accounting by interpreters. Grice established that linguistic meaning arises from a dialogic interplay of minds, along lines not dissimilar to the game-theoretic models of interpersonal strategies of communication and manipulation developed by economist Thomas Schelling (1960; see also Lewis 1969).

This breakthrough in the philosophical conception of how language works led to much subsequent research on the interpersonal logic of language use, which in turn drew attention to the kind of cognition that was required for language to be learned and used (see previous section). If language use requires us to be able to model the knowledge and intentions of our interlocutors in real time, then this means that the psychological wherewithal for language must include not only the capacity for symbol manipulation, but more importantly the higher-order social-cognitive capacities needed for "mindreading" (Sperber and Wilson 1986; Levinson 2000, 2006). In an example of how this line of work has been taken up in research on language evolution, Scott-Phillips (2015) emphasizes the role of the uniquely human cognitive advance of "ostensive-inferential communication," the essential insight of Grice (1957). Scott-Phillips states that humans are "adapted for ostensive communication, that is, to express and recognize communicative and informative intentions, and in doing so better navigate the social environment" (2015, 139).

But questions remain open. First, what does "navigating the social environment" mean? Scott-Phillips emphasizes the marked increase in group sizes that humans show, and notes that "large group sizes make for a very politicized existence" (2015, 140). But every group-living animal has a politicized existence. Are we talking about an existence that is merely "more" politicized than that of other primates, or was there a qualitative break? The answer, as we've demonstrated in this book, is that there was a qualitative break. Language-mediated micro-politics are more than mere matters of manipulation or influence over others. Language doesn't just allow us to navigate a social environment, it transforms and indeed constitutes the very social environments we must navigate. This is one of language's most momentous consequences.

Second, is the neo-Gricean emphasis on "mindreading" (something that listeners do) and "manipulation" (something that speakers do, exploiting what listeners do), perhaps paradoxically, overly individual-based? The account preserves a view of humans more as calculators than as agents. Scott-Phillips (2015, 141) is explicit about this: "communication is not a single trait, but rather the product of two other traits, namely mechanisms for signal production, and mechanisms for signal reception." That is, there are two individual-based mechanisms: "signallers try to mentally manipulate their audience, and audiences try to mindread their signallers" (ibid.). This notion of "social navigation" may require sophisticated cognitive abilities but it overly simplifies what interactants really do with language in interaction—under a tyranny of accountability—as we've endeavored to show in this book.

8.3 Consequences of Language

We have argued that language is grounded in an enchronic infrastructure for human interaction, and that this infrastructure contains a uniquely human form of primary intersubjectivity. In turn, we have argued that the infrastructure is a condition for a type of enhanced intersubjectivity that emerges through the operation and development of language within that infrastructure. Other species have complex interactions but they do not have culturally transmitted systems of signs akin to human language, with duality of patterning (in which semantic symbols are composed of distinctive nonsemantic units, allowing for unlimited growth of the lexico-semantic component), generative grammar (in which meaningful signs are combinable in patterns that are themselves meaningful), and the capacity for self-reference. Lacking these capacities has significant consequences for intersubjectivity.

Perhaps the most important feature of interaction sequencing in humans is that the structures it produces are accountable. If a recipient does not produce the conditionally relevant next action, this is an accountable fact, and people orient to it. Recipients may account for not responding (e.g., by indicating that their reason for not answering is that they do not know) and first-speakers may pursue such accounts if they are not offered. Another aspect of the accountability involved in interaction sequencing is seen in reports and complaints about third-party failures (e.g., "He didn't answer me"). Such reports radically transform the nature of human social life: "Humans alone among the social vertebrates can know one another substantially by repute" (Emler 1990, 177–178). Reputation has untold consequences for human sociality (see Boissevain 1974; Emler 1990; Enquist and Leimar 1993; Cohen 2012; Sperber and Baumard 2012; Origgi 2018). Other apes, not having available to them a flexible semiotic code like a human language, are accountable in the domain of brute forces, not social norms.

Philosophers have argued that human institutions would not exist without the reality-creating power of language as a device for enacting status-function declarations (Searle 2010). We argue that this power alone cannot account for how human institutional reality is created and maintained. Intersubjective understanding is necessary—as both condition and consequence—for any form of human activity. As we saw in the examples of sawing and weaving, it is not limited to overtly communicative activities. And this generalizes beyond face-to-face engagement.

All cooperative work involves an underlying enchronic architecture and so it makes little sense to isolate symbolic and communicative life from economic and material life. The enchronic infrastructure is a condition for all institutional reality, from markets to psychiatry, news interviews to court proceedings, medical consultation to healing ritual, crime to punishment, marriage to gift exchange. But these social realities are not consequences solely of the enchronic to-and-fro of intersubjective action. They have been made possible within that to-and-fro by the propositional, self-referential, and accountable systems of reference, semantics, and grammar. They are consequences of language.

Notes

Preface

1. See Hurford 2007, 2012; Levinson 2014; Scott-Phillips 2015; Tomasello 1999, 2008; and Everett 2017, among many others.

Chapter 1

1. Fortes (1983) made similar arguments about the dependence of society on rules and hence on language (see also Goody 1997). He writes: "what makes language specially significant is that it embodies most distinctly the function of culture that makes it the foundation of society. Culture is universally the vehicle of the constraint that is the essence of society; and language shows this paradigmatically because it exemplifies more rigorously and more visibly than any other constituent part of culture what I describe as the sovereignty of rules in social life" (4–5).

2. Others have expressed this insight. For example, Daniel Dor (2015, 4) writes: "First we invented language. Then language changed us."

3. And there is a kind of intersubjectivity that characterizes interactions between humans and other animals.

4. As we note below, turn-taking and action sequencing are indistinguishable in this context. Shared understanding of each move is not at issue here. In the sawing case, there are essentially two turns: (1) I pull while you push, and (2) you pull while I push. It's unlike talking in that the turns are simultaneous (though the complementary roles of turn-taking do overlap with the talking roles of speaker and listener). It's a binary system.

5. The numerals in the Lao transcriptions denote lexical tones (see Enfield 2007).

6. To *thematize* something is to place it in a common focus of attention to then characterize it in some way (see Kockelman 2007b). When we say *This tofu is amazing*, we are thematizing the tofu and characterizing it as amazing. Or we may thematize

the texture of the tofu, as in *This tofu is too firm*, or its taste, as in *This tofu is delicious*. The pragmatic relation between thematizing and characterizing is fundamental to the subject–predicate organization of linguistic propositions. While thematizing is a public phenomenon, it is ultimately grounded in cognitive organization: we attend to something in order to identify and evaluate it (Hurford 2003, 2007).

7. This and subsequent images are screen shots from our video recordings of reed-mat weaving sessions in Doune Ian Village, Vientiane Municipality, Laos. We are grateful to the participants for allowing us to record their work and publicize the research findings.

8. On the specificity principle in interaction, see Schegloff, Jefferson, and Sacks 1977; Clark and Shaefer 1987; and Dingemanse et al. 2015. The principle of choosing the more specific of a range of options when asking for clarification is both cooperative and efficiency-maximizing.

9. Durkheim uses the phrase "collective tyranny" in his book *Professional Ethics and Civic Morals* (see pages 61 and 64), though his better-known works do not mention "accountability" or "tyranny." Durkheim (1957, 61) writes: "When it is made up of a vast number of individuals, a society can exercise over each a supervision only as close and as vigilant and effective as when the surveillance is concentrated on a small number. A man is far freer in the midst of a throng than in a small coterie. Hence it follows that individual diversities can then more easily have play, that collective tyranny declines and that individualism establishes itself in fact, and that, with time, the fact becomes a right." This is not what we mean when we talk about a tyranny of accountability. Durkheim's meaning is more literal. He is pointing to the operation of centralized power and hinting at the possibility of the collective completely overwhelming the individual. Our notion of the tyranny of accountability foregrounds the orientation of individuals. People are not compelled to behave in such and such a way, but they are pervasively oriented to the potential consequences of any behavior.

10. Studies by Kidwell (e.g., 2013) and Kidwell and Zimmerman (2006) show that children as young as two years old design their actions with an eye to how they will be observed by others.

11. In broader usage, the words *norm* and *law* have much overlap. We do not use these English words with their full vernacular range. For example, the legal acts that embody policies of the European Union are sometimes referred to as norms. These are obviously not norms in the sense we mean here. The distinction between norms and laws is clear in principle, but as usual, in practice there is a gray area in which the distinction may be difficult to make. The word *law* is sometimes applied to forms of decision-making in nonliterate societies. Here, we distinguish *law* as a self-consciously separated domain from norms, with a body of texts that specialists consult to see if a wrong has been committed and, if so, what should be done about it. Norms, by contrast, are informal.

12. Technically speaking, a norm is a Peircean object in the sense of being recurrently pointed to by signs but never revealed as a thing in itself (Kockelman 2005; Enfield 2013, chap. 4). A norm has no existence apart from agents' orientations to it.

13. As scholars across a range of disciplines have shown, norms do not so much constrain as shape conduct. Thus, even when a person behaves in a way that might be seen to deviate from the norm, and so not be constrained by it, they typically, unless they are completely unaware, orient to that very norm, designing their conduct so as to convey this orientation. For instance, there is a norm against talking in movie theaters while the film is showing. This does not categorically prevent people from talking; rather, it encourages them to whisper when they do.

14. We have invented this word for the intended meaning (Latinate *ab-* "away from," *-liminal* "at a threshold"), as no appropriate word for our concept seems to exist.

15. We acknowledge of course that norms change, and that a crucial element of any such change is the very act of breaching the subliminal nature of the norm—that is, of holding people to account for a behavior that has so far been treated as entirely "normal." On the other hand, norms may dissolve over time (e.g., a man wearing a pink shirt or a ponytail).

16. *American Heritage Dictionary of the English Language*, 5th ed. (New York: Houghton Mifflin Harcourt, 2015).

17. https://patimes.org/quest-accountability/

18. US Department of Defence, *Dictionary of Military and Associated Terms*, 2005.

19. Our interest is the ongoing operation of these systems, as distinct from the (also important!) aspects of opening/closing the activities.

20. Let us clarify that we mean he is not doing anything else relevant to the activity of sawing. Of course, he may also be expressing frustration, accumulating merit, or paying a debt in some other interpretive frame.

Chapter 2

1. Alan Rappeport, "'What Is Aleppo?' Gary Johnson Asks in an Interview Stumble," *New York Times*, September 8, 2016, https://www.nytimes.com/2016/09/09/us/politics/gary-johnson-aleppo.html.

2. This doesn't mean that failing to meet expectations always removes one from a given status. One may "break the mold," as Donald Trump did in the same election.

3. Linton (1936, 115) distinguished between *ascribed* and *achieved* statuses: the former "are assigned to individuals without reference to their innate differences or abilities [and] can be predicted and trained for from the moment of birth"; the latter

are "those requiring special qualities" and are "left open to be filled through competition and individual effort."

4. Because interpretants are often publicly perceptible, they are, in turn, also signs that may be open to interpretation. The example of an interpretant we just gave was a person taking an umbrella on their way out the door, but in a subsequent frame this act could also be a *sign*. That is, if someone were to perceive it and respond to it, they might conclude that the actor has formed a belief that it is likely to rain while they are out.

5. This is unique to humans. While other animals are categorized with certain statuses—for example, a dog may be a "pet," a status with commonly defined expectations—their behavior is not plausibly guided by an awareness of, or any form of coordination around, that status. Most obviously, this is because a dog cannot invoke the status "pet" to explain or justify privileges that may come from their status (e.g., why the dog—but not the local birds or raccoons—should be allowed in the house during a storm).

6. Compare the similar distinction between parallel action and joint action, that is, the difference between (1) two strangers who happen to be walking side by side on a busy city street and (2) two people who are on a walk together (Gilbert 1990; Searle 2010). In the case of (2), a "public" agreement has been made, and can be referenced if needed.

7. Ben Cousins, "Couple Arrested after Spending US$120K Mistakenly Added to Their Account: Police," *CTV News*, September 9, 2019, https://www.ctvnews.ca/world/couple-arrested-after-spending-us-120k-mistakenly-added-to-their-account-police-1.4584829.

8. See work on semiotic epidemiology (Sperber 1985; Enfield 2003, 2014) and sociolinguistic network analysis (Milroy and Milroy 1992; Trudgill 2010). Statuses can be studied in these terms: just as sociolinguists map the distribution and variation of linguistic structures, so we should be able to map the distribution and variation of statuses.

9. These words—*rights*, *duties*, *entitlements*, *obligations*—have everyday meanings in English, and we acknowledge that they are steeped in the cultural history of the English language. To ensure that they have the culture-neutral sense that we intend, we define them here strictly in the following technical senses, using the terminology around the concepts of norms, introduced in chapter 1: (1) 'She has a *right/entitlement* to do X' means that X is *subliminal* (in the relevant community): Were she to do X (in a specified context), others would not be surprised or disposed to sanction her; (2) 'She has a *duty/obligation* to do X' means that X is *abliminal*: Were she NOT to do X (in a specified context), others WOULD be surprised and disposed to sanction her. With these technical senses, our use of these terms need not conform with the vernacular use of the words. For example, we will say that wearing clean shoes to work

in a downtown Toronto bank is both an entitlement (because by wearing clean shoes we do not elicit surprise or sanction) and an obligation (because by not wearing clean shoes we would elicit surprise and/or sanction). We will say that going to an Italian restaurant and ordering pizza is an entitlement, but not an obligation (because ordering pasta instead would be unremarkable).

10. Of course, status norms differ between communities and must be learned.

11. This is part of what is at issue in the debate about the status of police in contemporary Western societies, especially the United States. Specifically, questions have been raised as to whether police violence against African Americans can be attributed to the person as an occupant of the status police officer or instead imply divergence from the normative role requirements of police officers. This example also illustrates the significant ways in which normative expectations vary across groups. For instance, the expectation that police officers should give us assistance if we ask is clearly not unanimous in the United States. Indeed, there's substantial evidence to suggest that many African Americans are reluctant to call upon the police for fear that they or others may to come to harm as a result.

12. Uniforms are themselves graded. The example of the police officer's uniform represents one extreme, that of the banker's three-piece suit the other.

13. While ostensibly tied to biology in some way, these kinds of statuses show significant cultural variability. All cultures may distinguish statuses related to age bands, but they will pick out different bands than, say, those encoded in Anglo societies.

14. See also the practice of exchanging rings as part of a marriage ceremony. The ring, worn on a specified finger of the left hand, then becomes a uniform of the status "husband" or "wife."

15. New York State Penal Law, Article 190.26, "Criminal Impersonation in the First Degree." http://ypdcrime.com/penal.law/article190.htm#p190.26.

16. David WalleChinsky, "Is George Bush Guilty of War Crimes . . . and Who Cares?" *Huffpost*, The Blog, May 25, 2011, https://www.huffpost.com/entry/is-george-bush-guilty-of-_b_26669.

Chapter 3

1. Activities presuppose status but status does not presuppose activity. We can thus distinguish activity-derived statuses (e.g., speaker) from others (e.g., mother, woman).

2. From our perspective, statuses such as those of onlooker, bystander, audience member, and so on name kinds of participants, though this goes against common usage to some extent. These count as participant statuses because they entail engagement, albeit of a constrained type.

3. The distinction between what is announced and what is embodied was introduced in Schegloff and Sacks's article "Opening Up Closings" (1973). An essentially analogous distinction is made in linguistic anthropology between explicit and implicit metapragmatics or metapragmatic discourse and metapragmatic function (see, for example, Silverstein 1993).

4. Geertz incorporates a "sociology of culture" perspective in opening the chapter on the *prijaji* with a discussion of Robert Redfield and his distinction between great and little traditions. Geertz would later elaborate this point in *Negara: The Theatre State in Nineteenth-Century Bali* (1980): the cultural practices of a courtly elite constituted an exemplary center and thus a model for all. These trickled outwards from the center, dispersing and gradually dissipating the farther they traveled from court. Geertz also notes that influence can run in the opposite direction, as when elites, dissatisfied with the staid conservativism of the center, look to the periphery for inspiration in what they take to be more "raw," "living," or "authentic" forms of human expression.

5. Geertz's description of the *prijaji* worldview and practice is largely limited to men.

6. See also Siegel (1986, 18), who writes that when a child "speaks *Ngoko* to adults, his words are pushed aside and replaced with *Kromo*. Control of language means not the ability to say what one wants to say, but the ability to replace one's 'own' words, the words one would ordinarily use, with others. But these strange words honor others and, at the same time, give the child who speaks them a certain standing. He is on the way to becoming someone who can behave like an adult—in Javanese parlance, on the way to becoming 'Javanese,' the opposite of which is not another ethnic label but 'animal-ish.'"

7. See also Goffman (1963b) and Brooks-Gardner (1980) on "civil inattention," which can be seen as a way of respecting the individual autonomy or "information preserve" of nonintimates. Goffman (1963b, 84) writes: "What seems to be involved is that one gives to another enough visual notice to demonstrate that one appreciates that the other is present . . . , while at the next moment withdrawing one's attention from him so as to express that he does not constitute a target of special curiosity or design. In performing this courtesy the eyes of the looker may pass over the eyes of the other, but no 'recognition' is typically allowed. Where the courtesy is performed between two people passing on the street, civil inattention may take the special form of eyeing the other up to approximately eight feet . . . , and then casting the eyes down as the other passes. . . . In any case, we have here what is perhaps the slightest of interpersonal rituals, yet one that constantly regulates the social intercourse of persons in our society."

8. On similar ideas concerning negative forms of mutual orientation, see Hinde (1997) on enemies, and Scheflen (1976), Goffman (1963b), and Kendon (1990) on civil inattention.

9. In contexts in which a prohibition is against vocalizing rather than against communication per se, nonvocal forms of communication can develop, as if to

compensate. In Australian Aboriginal communities, norms of silence have provided a context for the evolution of auxiliary sign languages (Kendon 1988), not unlike the compensatory creation of alternative sign systems when the usual mode of participation is limited for practical reasons, for example, sawmill hand signs (used because vocal participation is constrained by noise; see Meissner, Philpott, and Philpott 1975a and b); hunting hand signs (used to prevent prey from being alerted by noise; see Mohr 2015); and semaphore, smoke signals, or "drum language" (used because vocal participation is constrained by distance; see Seifart et al. 2018).

10. The situation is similar among the Navajo (Mowrer 1970, cited in Basso 1970, 96–97).

11. We report these claims here, but we note that they can only be evaluated in light of the distinction between *ideologies* of participation—what people believe, or profess to believe, about participation in a given cultural context—and the *facts* of participation in real life (cf. the nonabsolute nature of the avoidance of interaction with one's brother-in-law in Guyana; see our discussion in chapter 2). Basso's investigation of silence among the Western Apache shows that silence is mandated in specific contexts, not across the board. Similarly, in the case of "the silent Finn," an impression that silence is everywhere may come from the fact that silence occurs in certain particularly salient contexts or moments (see also Bauman 1983).

12. Thompson ([1965] 1987, 305), in what remains the authoritative English-language reference grammar of Vietnamese, writes: "Distinctions between address and reference are seldom formal; which is involved is usually clear in the context. It can be generalized, however, that the demonstrative marker *ấy* quite often serves primarily to make clear that third person reference (and not second person address) is intended. Still clearer is the use of anaphorics . . . in southern and south central dialects. Nevertheless there are many cases where the forms . . . appear without any formal signal that they refer to a person other than the hearer or reader."

13. Ted Sammons, "'I Didn't Say Look; I Said Listen': The People's Microphone, #OWS, and Beyond," *Sounding Out!* https://soundstudiesblog.com/2011/11/21/i-didnt-say-look-i-said-listen-the-peoples-microphone-ows-and-beyond/.

14. Much of the literature on publics challenges not the idea(l) of the public sphere but its practical implementation. A public in which "individual interests" and "social statuses" are bracketed is, practically speaking, impossible. At the same time, any such public necessarily marginalizes if not excludes those who do not meet the requirements of participation.

15. Precisely this kind of reading public was the audience for Charles Darwin's 1859 book *On the Origin of Species by Means of Natural Selection*. See Horenstein (2009), Numbers and Stenhouse (1999), inter alia.

Chapter 4

1. Another set of contingencies that beset sinsigns are a function of the materiality of the signs themselves. As Peirce put it: "Since a sign is not identical with the thing signified, but differs from the latter in some respects, it must plainly have some characters which belong to it in itself, and have nothing to do with its representative function. These I call the material qualities of the sign" (1992, 40). This results in bundling effects (Keane 2003). For example, the redness of an apple is necessarily "bundled" with the roundness of the apple. Such bundles can give rise to collateral effects when one element of the bundle is selected and another element is unwittingly deployed (Sidnell and Enfield 2012).

2. This is the *tyranny of accountability*, which "provides us with the coordinates for seeing meaning in others' behavior, for planning our own behavior in anticipation of how others will see it, and thus for the design and interpretation of every next move in the progression of social life" (Enfield and Sidnell 2017a, xv). The idea is grounded in the insights of Garfinkel (1967) and Grice (1975): each member of a community finds themself "not so much constrained by rules or sanctions, as caught up in a web of inferences" (Levinson 1983, 321), inferences that "a person is never free from and never has the luxury of disregarding" (Enfield and Sidnell 2017, 55).

3. Levinson's neo-Gricean theory of conversational implicature was framed as a solution to this bottleneck problem. He asserted that while speakers can't go faster than the bottleneck allows, listeners can make inferences faster and so speakers will design their utterances to elicit such ampliative inferences and thereby dramatically increase the rate of information transfer (Levinson 2000). Thus, while Levinson starts with a speaker-only microgenetic problem (slow articulation), he introduces the relationship between speaker and hearer as the locus of the solution. The neo-Gricean theory of meaning is thus essentially enchronic in nature, as its unit of analysis is the sign-interpretant (and signifier-interpreter) *relation* within the context of a communicative sequence.

4. While talk-in-interaction is the paradigmatic case, enchronic relations are ubiquitous to human social and semiotic life. As Keane (1997, xiv) puts it: "Understood as 'practices,' representations do not exist only in the abstract—as, for example, some disembodied 'discourse.' Rather, that they take concrete forms, situated in activities, is critical to their signifying, performative, and even causal capacities."

5. The terms "complaining, inviting, requesting, agreeing, excusing, accepting, and granting" are used here for analytic convenience and do not imply any claims about the ontology of action in interaction (see Enfield and Sidnell 2014).

6. One aspect of this—the cultural aspect—is discussed below. Despite obvious differences between courtroom proceedings and news interviews (see Heritage and Clayman 2010), turn-taking always involves fundamental features, including projection of possible turn completion and linguistic techniques for turn construction.

7. The Kri live in small hamlets on the fringes of the nation-state in a remote section of the Nam Noi watershed in Laos (Enfield and Diffloth 2009; Enfield 2009; Zuckerman and Enfield 2020, 2022; Enfield and Zuckerman in press). Kri belongs to the Vietic subbranch of Austroasiatic and is currently spoken by around 500 people.

8. Moreover, the talk is accompanied by a gaze-shift point that reaches its apex on *meengq* (presumably indicating the location of the "others"), such that by the time Sii produces the final word here, *qo'o'k*, his gaze, having briefly caught the addressee on its return journey, is now focused on the knife he is using to cut wood (on gaze and turn transition, see Goodwin 1980).

9. We have only scratched the surface of research on the organization of interaction. The manifold ways in which a response displays an understanding of the talk to which it responds have been described in studies by, for instance, Button (1992), Ehrlich and Sidnell (2006), Enfield and Sidnell (2015), Gardner (2001), Goffman (1976), Goodwin (1986), Heritage (1984b, 1998, 2002), Heritage and Sefi (1992), Jefferson (1980), Pomerantz (1978, 1984), Raymond (2003), Sacks ([1973] 1987, 1975), Schegloff (1968, 1982, 1988b, 1988c, 1992, 1996), Sidnell (2007), Stivers (2005), Thompson, Fox, and Couper-Kuhlen (2015), Wilkinson and Kitzinger (2006), and Wootton (1997).

10. Gesture isn't the only source of contrast to the turn-organized nature of interaction. Not all aspects of talk contribute to the constraints on turn-taking that we describe here. Many features of prosody can be controlled independently of lexical and syntactic choices, and can create or extend structures either within or beyond the limits of the turn.

11. Such understandings are seldom if ever perfect in the sense of an exact match between what a speaker intended and what a recipient understood. Intention in language does not exist independently of the contexts in which it is made manifest. As Wittgenstein once said, you can't intend to put your opponent in check if the game of chess does not exist. In interaction, someone's understanding of an action should be *adequate*, as measured by that person's behavior in response to the action (see Garfinkel 1967). In interaction, understanding is not "perfect" but *good enough*.

12. We draw here on the analysis by Raymond and Sidnell (2014).

Chapter 5

1. The argument that reference is *necessarily* social has been made many times in many ways. Wittgenstein argued against the notion of a private language on the grounds that if the language weren't shared there would be no way of knowing whether a particular use was correct, or, as he put it, "thinking one was obeying a rule would be the same thing as obeying it" (1953, §202).

2. We note that there are exceptional cases in which gaze is not directed to the intended referent (see Cooperrider et al. 2021).

3. Finger-pointing (Kita 2003) also allows for a higher informational load than do the forms of lip/gaze-pointing we have considered. Pointing practices have developed a range of functional contrasts associated with variance in the "size" of the gesture (Enfield, Kita, and de Ruiter 2007; Kita and Essegbey 2001; Kita 2003; Cooperrider et al. 2021), whether the gesture is accompanied by gaze (Streeck 1993) or which hand or finger configuration is used (Wilkins 2003; Kendon and Versante 2003). Finger-pointing also makes path descriptions possible, as well as illustrative combinations. At the far end of the informational scale are diagrammatic representations in which pointing gestures are used to identify positions within a virtual drawing (see Enfield 2005).

4. In this respect note that lip-pointing—which crucially involves gaze as noted above—seems specialized among Lao speakers in two ways: First, "lip-pointing is apparently restricted to cases when the addressee is looking at the speaker" (Enfield 2001, 192) and second to "acts of direct ostension in which the location or identity of a referent in the physical environment is in focus" (Enfield 2001, 196). The prior establishment of recipiency along with the already in focus character of the referent obviates or at least alleviates the need to monitor the recipient.

5. On the one hand, there are many words in any language that seem to rely for their meaning not just on a whole host of other words to which they are opposed but also on practices of metasemantic glossing. For instance, abstract words such as *justice* and *wisdom* could hardly have the meanings they have if there were no way to explicate their meanings. Deacon (1997) suggests that the dependency of words on other words by which they are explicable is not limited to such abstract notions but is generally characteristic of symbols. See also Schegloff (1989, 1991) on the dependency of human languages on practices of repair, which are by definition metalinguistic.

6. Language-encoded information is fundamentally digital and so is available for analysis in terms of units that are essential to its own organization (rather than imposed from without). There are two key differences from embodied action. First, embodied action is, at every point, tied to a spatial and artifactual context; for example, a person designs the action of getting up from the table by taking into account the position of the chair, the table, the other persons close by, etc. Second, embodied action's organization is largely analogue rather than digital. This is seen in the way one component links to the others (movement of the hand, arm, shoulder, upper body, etc.) with no clear distinctions between components and in terms of the action sequence, that is, each component action within a sequence bleeds into the next. The result is that the sentence/proposition "He got up quickly from the table without saying a word" is much more amenable to analysis and thus accountability than is the actual behavior described (his way of getting up from the table). When there is a question of *what it is about* the way he got up, that question will be settled by words. Note also that there are issues involved in the fact that any nonverbal depiction (like that which it depicts) will be tied to the interactive setting in which *it* is produced. It is also indexical through and through.

7. For discussion of some of the complexities, see, *inter alia*, Drew (1997) and Robinson (2006, 2014).

8. The term "framing" here refers to the provision of a structural/grammatical frame for some referred-to speech.

Chapter 6

1. Such evaluation is not done in an active or conscious manner, at least among fully socialized members of a group. The conformity of other people's interpretants is abliminal. We only attend to the matter of agreement in understanding when we detect some kind of *mismatch*. In the nature of normative organization, the question of whether others converge in their understanding of shared systems of communication is *abliminal*. We only attend to it when it goes counter to expectation (or more accurately, subprehension).

2. It might be possible to distinguish between pragmatic norms (regulating the metalingual or procedural functions of language) and semantic norms (regulating the referential function of language), if they are taken as two aspects of a synchronic system of social norms that surround communication.

3. "You and I can sensibly discuss how many books I have on my shelf, for example, only if we both agree, at least roughly, about what a book is. We can disagree over borderline cases: I may call something a slim book that you would call a pamphlet. But we cannot disagree over what I called pivotal cases. If you do not count my copy of *Moby-Dick* as a book because in your view novels are not books, any disagreement is bound to be senseless." (Dworkin 1986, 45) Dworkin's point about questions of law is fundamentally a question of language. We would add that not only would a disagreement over pivotal cases be "senseless" (in our earlier example, it might be somebody saying that apples aren't fruit), but it would be grounds to sanction somebody as foreign, deviant, or intentionally uncooperative. Agreeing on what pivotal cases mean is not optional. You could deny that *Moby-Dick* is a book, but you could not *plausibly* deny it. Thus, Dworkin's point about pivotal cases is not just about being sensible, but being an accountable community member.

4. We are not claiming that when Virginia challenges Mom's use of the word "child" she is motivated by a concern for the semantics of English (though such a concern might apply in other contexts). Her motivations in thematizing the applicability of a word to a referent are orthogonal to the normative basis that she assumes in making that action.

5. "And only one for birthday presents, you know. There's glory for you!"

"I don't know what you mean by 'glory,'" Alice said.

Humpty Dumpty smiled contemptuously. "Of course you don't—till I tell you. I meant 'there's a nice knock-down argument for you!'"

"But 'glory' doesn't mean 'a nice knock-down argument,'" Alice objected.

> "When I use a word," Humpty Dumpty said, in rather a scornful tone, "it means just what I choose it to mean—neither more nor less."
>
> "The question is," said Alice, "whether you can make words mean so many different things."
>
> "The question is," said Humpty Dumpty, "which is to be master—that's all." (Carroll [1871] 2009, 190)

6. See Ehrlich and Sidnell (2006) and Stivers and Hayashi (2010) for more examples along these lines.

7. We also know that linguistic description can affect the psychology of perception (or our conception of perception) and reasoning (see Enfield 2022, chap. 4). For example, when a single scene is described to one person as a car "hitting" another car versus "crashing into" another car, the two people's estimations of the speed of the crash will be different (Loftus and Palmer 1974). And when a scene is described agentively (*She toppled the candlestick and then the fire started*) versus nonagentively (*The candlestick toppled and then the fire started*), this can affect people's judgements of culpability (Fausey and Boroditsky 2010).

8. This is a case in which *status-based authority* outranks *source-based authority* (Enfield 2013).

9. See also Raymond (2019) for a discussion of cases in which a speaker anticipates a normatively expectable inference of what they have said and preemptively defeases it: for example, the speaker characterizes a bike as "not a mountain bike" and subsequently announces, "did go up a mountain on it."

10. Prudence may have reasons for disclaiming knowledge. In doing so, she aligns with her mother-in-law and distances herself from the youth culture in which the word *gwaff* presumably circulates, thereby appearing more mature than she might otherwise.

11. Such mental states cannot exist independently of their representation, even if only their representation to the one who holds them. Part of what "believing something" means is that that something is represented by and to the one who holds the belief as true. In a semiotic account, this is no less the case with other intentional mental states of *hoping*, *wishing*, *fearing*, and so on (see also Kockelman 2013).

12. The asterisk at the beginning of the second example indicates that it is "ungrammatical," meaning that it is not something people would say, or at least people would judge it as somehow "wrong."

Chapter 7

1. Grammar also includes the sound system of a language in its various phonological rules and subsystems, but these are not the focus of this chapter.

2. We do not want to imply that a distinction between lexicon and grammar is clear cut. We use it here as a practical heuristic for thinking about the division of labor

among the resources of language. We agree with analysts such as Ron Langacker, who has argued that there is no essential or qualitative distinction between lexicon, morphology, syntax, or idiom: all involve semantic coding with combinatoric potential (Langacker 1987; Wierzbicka 1988; Croft 2001; Goldberg 2006).

3. Terms such as theme, rheme, subject, and predicate have been endlessly contested and redefined in the history of the language sciences, especially through the twentieth century.

4. A key element of the ideas of "on the record" and "in the open" is the principle of *undeniability*: if it is in the common ground then one cannot deny that one knows it, that others know it, and, critically, that one knows others know that one knows.

5. The infinite potential for expression thanks to the generative nature of language has of course been noted since Humboldt ([1836] 1988; see also Chomsky 1957; Adger 2019 inter alia), but seldom with reference to the social implications of this generativity.

6. Some have claimed that certain limited types of animal communication do show (at least partial) displacement. For example, bees use a waggle dance to communicate the location of a food source to other bees in the hive.

7. For a survey of many of these, under the rubric of "engagement," see Evans et al. 2018a and 2018b. Engagement is defined, following Landaburu (2007), as "a grammatical system for encoding the relative accessibility of an entity or state of affairs to the speaker and addressee" (Evans et al. 2018a, 118).

8. "Aspect" covers diverse meanings. Some aspectual marking is purely concerned with conveying information about a narrated event, for example whether an action was momentary or repeated. But much aspectual marking also incorporates meaning relating to an event's temporal/situational relevance to the speech event. So, *Kim has fallen asleep*, in the present perfect, conveys that the event of falling asleep is completed (relating to En), but also that its result—Kim's being in a sleeping state—persists now (relating to Es).

9. Epistemic modality (e.g., *Kim must be asleep*) links En to Es in that it denotes the speaker's degree of certainty or commitment to the truth of the proposition (i.e., something compels *the speaker*—in Es—to believe that Kim is asleep). By contrast, deontic modality is confined to En: *Kim must go to sleep* (in which something compels *Kim*—in En—to go to sleep).

10. Certain grammatical categories such as some case-marking on nouns and some aspectual marking on verbs exclusively convey information about En, without reference to Es.

11. See examples below showing that an assertion is understood as "an answer" when it occurs immediately after somebody's question. And consider cases like *Huh?* which must be understood as orienting directly to "what you just said" (see our

discussion of *Huh?* as a verbal point in chapter 5). Also see Clark's (1996) claim that every utterance carries with it certain reflexive collateral questions such as "Have you understood what I just said?" and "Can I continue?", accounting for why "continuers" are often also words that mean "Yes" in response to explicit questions.

12. Recall the example given by Harvey Sacks (1972) from a children's book: *The baby cried. The mommy picked it up.* From these two propositions in sequence, we understand, among other things, that the mommy is the mommy of the baby, that she picked the baby up immediately after it cried, that she did so in order to comfort it, and so on.

13. For another example, see Sarvasy (2020) on the Nungon language of Papua New Guinea. Here is the English rendering of one of Sarvasy's examples of a Nungon clause-chaining construction: "The sun being$_{\text{medial}}$ [there], we seeing$_{\text{medial}}$ [the sun] in the evening, the weather being$_{\text{medial}}$ fine, readying$_{\text{medial}}$ our food for the journey, packing$_{\text{medial}}$ it, readying$_{\text{medial}}$ and packing$_{\text{medial}}$ our bows and arrows, fastening$_{\text{medial}}$ our bowstrings, we slept$_{\text{final}}$." Note that in Nungon there are same-subject and different-subject markers for medial verbs, where same-subject markers have zero grammatical information, and different-subject markers inflect for person/number.

14. Some exceptions have been claimed (e.g., Everett 2005; cf. Hale 1976; Givón 1979, 306; Foley 1986, 177; Dixon 1995). But even without unambiguous grammatical marking of clausal embedding, language users faced with an utterance translated as, say, "he say his canoe capsize" will readily distinguish between interpretations such as "He said (something), (and) his canoe capsized" versus "He said that his canoe capsized"; that is, language users will have not trouble with the idea that one person can refer to the content of another person's speech.

15. See Wierzbicka (1988, chap. 1) for a detailed explication of the semantics of English complementization structures.

16. The central argument, versions of which may be found not only in Wittgenstein's remarks on the foundations of psychology and Ryle's penetrating critique in *The Concept of Mind* (1949) but also in works of Vygotsky and many others, is that any adequate account of human cognition must acknowledge the critical role played by human representations of and rationalizations *about* cognition, since for humans there are no purely individual psychological processes independent of intersubjective processes.

17. See *The World Atlas of Language Structures Online*, "Relationship between the Order of Object and Verb and the Order of Adposition and Noun Phrase," n.d., http://wals.info/combinations/95A#2/15.0/153.0.

18. Arguments in the microgenetic frame also often implicitly invoke the *ontogenetic* frame, highlighting the effects of the child's capacity for, and strategies in, the learning of grammatical structure during the life course. Principles of human cognitive

development in ontogeny impose biases on how language is acquired, which in turn affects languages in the longer term, in diachrony (see Bowerman and Levinson 2001; Clark 2003; Trudgill 2011).

19. See discussion and references in Martin and Culbertson (2020), whose study shows that speakers of a heavily prefixing language—Kĩĩtharaka, a Central Kenya Bantu language—do not behave like speakers of a suffixing language in a nonlinguistic perception task that taps into a preference for front-loading information in linguistic structures. Martin and Culbertson (2020) speculate that speakers of Kĩĩtharaka have rewired their perceptual preferences in accordance with the fine-grained linear organization of information in their language.

20. If the momentum of interaction is toward suffixes, what accounts for the 16 percent of languages that prefer prefixes? The answer is that interactional forces are not direct determinants of linguistic structure but are sources of bias within systems that are subject to a vast array of other, diverse biases. Language evolution—like biological evolution—is the outcome of a vast (but finite) set of events, so enormous and varied in its possibilities that outcomes can never be known in advance or predicted with complete accuracy.

Chapter 8

1. "We who are grammatically trained may use a simple ending to correct an idea previously expressed. Thus the statement *He sings beautifully* might elicit a reply, *sang*; or a laconically inclined person might even remark, in reply to the statement *He plays well*, *-ed*, which by his friends might be well understood" (Boas 1911, 28).

2. Principles of semiotic motivation, equivalent in nature to those that underlie the bodily expressions of emotion as analyzed by Darwin (1872), are likely to have regulated the formal development of such interjections, as they clearly still do today (see Ohala 1984, 1994; Enfield et al. 2013; Dingemanse, Torreira, and Enfield 2013).

3. This idea is found in diverse proposals made by fellow travelers including Tomasello (2008), Hurford (2008, 2012), Levinson (2014), inter alia.

4. Accordingly, subsequent empirical work comparing human infants and other great apes finds that young humans are not much cleverer than other primates in *non*-social cognitive domains (space, quantities, causality) but they are much cleverer in aspects of *social* cognition (social learning, theory of mind). See Herrmann et al. (2007). and Tomasello and Herrmann (2010).

5. Crucially, when we say "language as we know it," we include not just the "design features" proposed by linguists since Hockett, but the conversational structures of turn-taking, sequence organization, repair, etc.

References

Adger, David. 2019. *Language Unlimited: The Science behind Our Most Creative Power*. Oxford: Oxford University Press.

Agha, Asif. 2007. *Language and Social Relations*. Cambridge, UK: Cambridge University Press.

Anscombe, G. E. M. 1957. *Intention*. Oxford: Blackwell.

Anscombe, G. E. M. 1979. "Under a Description." *Noûs* 13, no. 2: 219-233.

Arcadi, A. C. 2000. "Vocal Responsiveness in Male Wild Chimpanzees: Implications for the Evolution of Language." *Journal of Human Evolution* 39, no. 20: 205–223.

Austin, John L. 1962. *How to Do Things with Words*. Oxford: Clarendon Press.

Baranova, Julija. 2015. "Other-Initiated Repair in Russian." *Open Linguistics* 1, no. 1: 555–577.

Baranova, Julija, and Mark Dingemanse. 2016. "Reasons for Requests." *Discourse Studies* 18, no. 6: 641–675.

Basso, Keith. 1970. "To Give Up on Words: Silence in Western Apache Culture." *Southwestern Journal of Anthropology* 26, no. 3: 213–230.

Bateson, Gregory. 1979. *Mind and Nature*. New York: E. P. Dutton.

Bauman, Richard. 1983. *Let Your Words Be Few: Symbolism of Speaking and Silence among Seventeenth-Century Quakers*. Cambridge, UK: Cambridge University Press.

Benveniste, E. 1966. "The Nature of Pronouns." In *Problems in General Linguistics*, translated by Mary Elizabeth Meek, 217–222. Coral Gables: University of Miami.

Bloch, M. 1975. "Introduction." In *Political Language and Oratory in Traditional Society*, edited by M. Bloch, 1–28. New York: Academic.

Bloomfield, Leonard. 1933. *Language*. Chicago: University of Chicago Press.

Bloomfield, Leonard. [1943] 1970. "Meaning." In *A Leonard Bloomfield Anthology*, edited by Charles Hockett, 400–405. Bloomington: Indiana University Press.

Boas, Franz. 1911. "Introduction." In *The Handbook of American Indian Languages*, 1–79. Washington, DC: Government Printing Office.

Boissevain, Jeremy. 1974. *Friends of Friends: Networks, Manipulators, and Coalitions*. New York: St Martin's Press.

Bourdieu, Pierre. 1977. *Outline of a Theory of Practice*. Cambridge, UK: Cambridge University Press.

Bowerman, Melissa, and Stephen C. Levinson. 2001. *Language Acquisition and Conceptual Development*. Cambridge, UK: Cambridge University Press.

Bratman, Michael E. 2007. *Structures of Agency*. Oxford: Oxford University Press.

Brooks-Gardner, Carol. 1980. "Passing By: Street Remarks, Address Rights, and the Urban Female." *Sociological Inquiry* 50: 328–356.

Brown, Penelope, and Suzanne Gaskins. 2014. "Language Acquisition and Language Socialization." In *The Cambridge Handbook of Linguistic Anthropology*, edited by N. J. Enfield, Paul Kockelman, and Jack Sidnell, 187–226. Cambridge, UK: Cambridge University Press.

Brown, Roger. 1958a. "How Shall a Thing Be Called?" *Psychological Review* 65, no. 1: 14–21.

Brown, Roger. 1958b. *Words and Things*. Glencoe, IL: The Free Press.

Brown, Roger, and Marguerite Ford. 1961. "Address in American English." *Journal of Abnormal and Social Psychology* 62, no. 2: 375–385.

Brown, Roger, and Albert Gilman. 1960. "The Pronouns of Power and Solidarity." In *Style in Language*, edited by T. A. Sebeok, 253–276. Cambridge, MA: MIT Press.

Bruner, Jerome. 1991. "The Narrative Construction of Reality." *Critical Inquiry* 18, no. 1: 1–21.

Bühler, K. [1934] 1982. The deictic field of language and deictic words, in *Speech, place, and action*, edited by R. J. Jarvella and W. Klein, 9–30. New York: Wiley.

Bunnag, Jane. 1973. *Buddhist Monk, Buddhist Layman: A Study of Urban Monastic Organization in Central Thailand*. Cambridge, UK: Cambridge University Press.

Burridge, Kenelm. 1975. "The Melanesian Manager." In *Studies in Social Anthropology: Essays in Memory of E. E. Evans-Pritchard by His Former Oxford Colleagues*, edited by John H. M. Beattie and R. G. Lienhardt, 86–104. Oxford: Clarendon Press.

Button, Graham. 1992. "Answers as Interactional Products: Two Sequential Practices Used in Interviews." In *Talk at Work*, edited by P. Drew and J. Heritage, 212–231. Cambridge, UK: Cambridge University Pres.

Bybee, Joan. 2010. *Language, Usage, and Cognition*. Cambridge, UK: Cambridge University Press.

Byrne, Richard, and Andrew Whiten. 1978. *Machiavellian Intelligence: Social Expertise and the Evolution of Intellect in Monkeys, Apes, and Humans*. Oxford: Clarendon Press.

Carpenter, Malinda, Katherine Nagell, Michael Tomasello, George Butterworth, and Chris Moore. 1988. "Social Cognition, Joint Attention, and Communicative Competence from 9 to 15 Months of Age." *Monographs of the Society for Research in Child Development* 63, no. 4.

Carroll, Lewis. [1871] 2009. *Alice's Adventures in Wonderland and Through the Looking-Glass*. Oxford: Oxford University Press.

Cheney, Dorothy L., and Robert M. Seyfarth. 1992. *How Monkeys See the World: Inside the Mind of Another Species*. Chicago: University of Chicago Press.

Chomsky, Noam. 1957. *Syntactic Structures*. The Hague: Mouton and Co.

Christiansen, Morten H., and Nick Chater. 2008. "Language as Shaped by the Brain." *Behavioral and Brain Sciences* 31: 489–558.

Clark, Eve V. 1999. "Acquisition in the Course of Conversation." *Studies in the Linguistic Sciences* 29, no. 2: 1–18.

Clark, Eve V. 2003. *First Language Acquisition*. Cambridge, UK: Cambridge University Press.

Clark, Herbert H. 1996. *Using Language*. Cambridge, UK: Cambridge University Press.

Clark, Herbert H., and E. F. Schaefer. 1987. "Collaborating on Contributions to Conversations." *Language and Cognitive Processes* 2, no. 1: 19–41.

Clark, Herbert H., Robert Schreuder, and Samuel Buttrick. 1983. "Common Ground at the Understanding of Demonstrative Reference." *Journal of Verbal Learning and Verbal Behavior* 22: 245–58.

Clay, Zanna, and Klaus Zuberbühler. 2011. "Bonobos Extract Meaning from Call Sequences." *PLoS ONE* 6, no. 4.

Cody, Francis. 2011. "Publics and Politics." *Annual Review of Anthropology* 40: 37–52.

Cohen, Emma. 2012. "The Evolution of Tag-Based Cooperation in Humans: The Case for Accent." *Current Anthropology* 53, no. 5: 588–616.

Cooperrider, Kensy, Jordan Fenlon, Jonathan Keane, Diane Brentari, and Susan Goldin-Meadow. 2021. "How Pointing Is Integrated into Language: Evidence from Speakers and Signers." *Frontiers in Communication* 6.

Croft, William. 2001. *Radical Construction Grammar: Syntactic Theory in Typological Perspective*. Oxford: Oxford University Press.

Cutler, Anne. 2012. *Native Listening: Language Experience and the Recognition of Spoken Words*. Cambridge, MA: MIT Press.

Dancygier, Barbara. 2012. *The Language of Stories: A Cognitive Approach*. Cambridge: Cambridge University Press.

Darwin, Charles. 1859. *The Origin of Species*. Oxford: Oxford University Press.

Darwin, Charles. 1872. *The Expression of the Emotions in Man and Animals*. London: John Murray.

Deacon, Terrence W. 1997. *The Symbolic Species: The Co-Evolution of Language and the Brain*. New York: W. W. Norton.

de Pina-Cabral, João. 2018. "Modes of Participation." *Anthropological Theory* 18, no. 4: 435–455.

de Ruiter, Jan-Peter, Holger Mitterer, and N. J. Enfield. 2006. "Projecting the End of a Speaker's Turn: A Cognitive Cornerstone of Conversation." *Language* 82, no. 3: 515–535.

de Saussure, Ferdinand. 1916. *Cours de Linguistique Générale*. Paris: Payot.

de Villiers, Jill. 2007. "The Interface of Language and Theory of Mind." *Lingua* 117, no. 11: 1858–1878.

de Villiers, Jill, and Peter A. de Villiers. 2014. "The Role of Language in Theory of Mind Development." *Topics in Language Disorders* 34, no. 4: 313–328.

Dingemanse, Mark. 2015. "Other-Initiated Repair in Siwu." *Open Linguistics* 1: 232–55.

Dingemanse, Mark, Joel Blythe, and Tyko Dirksmeyer. 2014. "Formats for Other-Initiation of Repair across Languages: An Exercise in Pragmatic Typology." *Studies in Language* 38, no. 1: 5–43.

Dingemanse, Mark, and N. J. Enfield. 2015. "Other-Initiated Repair across Languages: Towards a Typology of Conversational Structures." *Open Linguistics* 1: 96–118.

Dingemanse, Mark, and Simeon Floyd. 2014. "Conversation across Cultures." In *The Cambridge Handbook of Linguistic Antropology*, edited by N. J. Enfield, Paul Kockelman, and Jack Sidnell, 447–480. Cambridge, UK: Cambridge University Press.

Dingemanse, Mark, Sean G. Roberts, Julija Baranova, Joe Blythe, Paul Drew, Simeon Floyd, Rosa S. Gisladottir, Kobin H. Kendrick, Stephen C. Levinson, Elizabeth Manrique, Giovanni Rossi, and N. J. Enfield. 2015. "Universal Principles in the Repair of Communication Problems." *PLoS ONE* 10, no. 9.

Dingemanse, Mark, Francisco Torreira, and N. J. Enfield. 2013. "Is 'Huh?' a Universal Word? Conversational Infrastructure and the Convergent Evolution of Linguistic Items." *PLoS ONE* 8, no. 11.

Dixon, R. M. W. 1995. "Complement clauses and complementation strategies." In *Grammar and Meaning: Essays in Honor of Sir John Lyons*, edited by F. R. Palmer, 175–220. Cambridge: Cambridge University Press

Dixon, R. M. W. 2002. *Australian Languages*. Cambridge, UK: Cambridge University Press.

Dixon, R. M. W. 2005. *A Semantic Approach to English Grammar*. Oxford: Oxford University Press.

Dixon, R. M. W. 2010. *Basic Linguistic Theory*. Oxford: Oxford University Press.

Dixon, R. M. W. 2014. "The Basics of a Language." In *The Cambridge Handbook of Linguistic Anthropology*, edited by N. J. Enfield, Paul Kockelman, and Jack Sidnell, 29–47. Cambridge, UK: Cambridge University Press.

Dixon, R. M. W., and Alexandra Y. Aikhenvald, eds. 2006. *Complementation: A Cross-Linguistic Typology*. New York: Oxford University Press.

Dor, Daniel. 2015. *The Instruction of Imagination: Language as a Social Communication Technology*. Oxford: Oxford University Press.

Drew, Paul. 1981. "Adults' Corrections of Children's Mistakes." In *Adult-Child Conversations*, edited by Peter French and Margaret MacLure, 244–267. London: Croom Helm.

Drew, Paul. 1992. "Contested Evidence in Courtroom Examination: The Case of a Trial for Rape." In *Talk at Work: Interaction in Institutional Settings*, edited by Paul Drew and John Heritage, 470–520. Cambridge, UK: Cambridge University Press.

Drew, Paul. 1997. "'Open' Class Repair Initiators in Response to Sequential Sources of Trouble in Conversation." *Journal of Pragmatics* 28: 69–101.

Dryer, M. S. 2013. "Prefixing vs. Suffixing in Inflectional Morphology." In *The World Atlas of Language Structures Online*, edited by M. S. Dryer and M. Haspelmath. Leipzig: Max Planck Institute for Evolutionary Anthropology.

Dunbar, Robin I. M. 1985. *Reproductive Decisions: An Economic Analysis of Gelada Baboon Social Strategies*. Princeton, NJ: Princeton University Press.

Dunbar, Robin I. M. 1993. "The Co-Evolution of Neocortical Size, Group Size, and Language in Humans." *Behavioral and Brain Sciences* 16, no. 4: 681–694.

Dunbar, Robin I. M., N. D. C. Duncan, and Anna Marriot. 1997. "Human Conversational Behavior." *Human Nature* 8, no. 3: 231–246.

Dunbar, Robin I. M. 2003. "The Social Brain: Mind, Language, and Society in Evolutionary Perspective." *Annual Review of Anthropology* 32: 163–81.

Duncker, Dorthe. 2019. *The Reflexivity of Language and Linguistic Inquiry: Integrational Linguistics in Practice*. London and New York: Routledge.

Durkheim, Emile. 1957. *Professional Ethics and Civic Morals*. London and New York: Routledge.

Dworkin, Ronald. 1986. *Law's Empire*. Cambridge, Mass: Belknap Press.

Egbert, Maria, Andrea Golato, and Jeffrey D. Robinson. 2009. "Repairing Reference." In *Conversation Analysis: Comparative Perspectives*, edited by Jack Sidnell, 104–132. Cambridge, UK: Cambridge University Press.

Ehrlich, Susan, and Jack Sidnell. 2006. "'I Think That's Not an Assumption You Ought to Make': Challenging Presuppositions in Inquiry Testimony." *Language in Society* 35: 655–676.

Emler, Nicholas. 1990. "A Social Psychology of Reputation." *European Review of Social Psychology, Vol. 1*, edited by Wolfgang Stroebe and Miles Hewstone, 171–193. New York: John Wiley and Sons.

Enfield, N. J. 2001. "'Lip-Pointing'—A Discussion of Form and Function with Reference to Data from Laos." *Gesture* 1, no. 2: 185–212.

Enfield, N. J. 2003. "Demonstratives in Space and Interaction: Data from Lao Speakers and Implications for Semantic Analysis." *Language* 79, no. 1: 82–117.

Enfield, N. J. 2005. "The Body as a Cognitive Artifact in Kinship Representations. Hand Gesture Diagrams by Speakers of Lao." *Current Anthropology* 46, no. 1: 51–81.

Enfield, N. J. 2006. "Social Consequences of Common Ground." In *Roots of Human Sociality: Culture, Cognition, and Interaction*, edited by N. J. Enfield and Stephen C. Levinson, 399–430. London: Berg.

Enfield, N. J. 2007. *A Grammar of Lao*. Berlin: Mouton de Gruyter.

Enfield, N. J. 2009a. *The Anatomy of Meaning: Speech Gesture, and Composite Utterances*. Cambridge, UK: Cambridge University Press.

Enfield, N. J. 2009b. "Everyday Ritual in the Residential World." In *Ritual Communication*, edited by Gunter Senft and Ellen B. Basso, 51–80. Oxford: Berg.

Enfield, N. J. 2013. *Relationship Thinking: Agency, Enchrony, and Human Sociality*. New York: Oxford University Press.

Enfield, N. J. 2014. "The Item-System Problem." In *The Cambridge Handbook of Linguistic Anthropology*, edited by N. J. Enfield, Paul Kockleman, and Jack Sidnell, 48–77. Cambridge, UK: Cambridge University Press.

Enfield, N. J. 2015. *The Utility of Meaning: What Words Mean and Why* (1st ed.). Oxford Linguistics. Oxford: Oxford University Press.

Enfield, N. J. 2017. *How We Talk: The Inner Workings of Conversation*. New York: Basic Books.

Enfield, N. J. 2021. *The Languages of Mainland Southeast Asia*. New York: Cambridge University Press.

Enfield, N. J. 2022. *Language vs. Reality: Why Language is Good for Lawyers and Bad for Scientists*. Cambridge, MA: MIT Press.

Enfield, N. J. In press. "Enchrony". *WIRES Cognitive Science.*

Enfield, N. J., and Gérard Diffloth. 2009. "Phonology and Sketch Grammar of Kri, a Vietic Language of Laos." *Cahiers de Linguistique—Asie Orientale* 38, no. 1: 3–69.

Enfield, N. J., Mark Dingemanse, Julija Baranova, Joe Blythe, Penelope Brown, Tyko Dirksmeyer, Paul Drew, Simeon Floyd, Sonja Gipper, Rosa S. Gisladottir, Gertie Hoymann, Gertie, Kobin H. Kendrick, Stephen C. Levinson, Lilla Magyari, Elizabeth Manrique, Giovanni Rossi, Lila San Roque, and Francisco Torreira. 2013. "Huh? What?—A First Survey in 21 Languages." In *Conversational Repair and Human Understanding,* edited by Makoto Hayashi, Geoffrey Raymond, and Jack Sidnell, 343–380. Cambridge, UK: Cambridge University Press.

Enfield, N. J., Sotaro Kita, and Jan-Peter de Ruiter. 2007. "Primary and Secondary Pragmatic Functions of Pointing Gestures." *Journal of Pragmatics* 39, no. 10: 1722–1741.

Enfield, N. J., and Paul Kockelman, eds. 2017. *Distributed Agency.* Oxford: Oxford University Press.

Enfield, N. J., and Stephen C. Levinson, eds. 2006. *Roots of Human Sociality: Culture, Cognition, and Interaction.* London: Routledge.

Enfield, N. J., and Jack Sidnell. 2014. "Language Presupposes an Enchronic Infrastructure for Social Interaction." In *The Social Origins of Language: Studies in the Evolution of Language,* edited by Daniel Dor, Chris Knight, and Jerome Lewis, 92–104. Oxford: Oxford University Press.

Enfield, N. J., and Jack Sidnell. 2015. "Language Structure and Social Agency: Confirming Polar Questions in Conversation." *Linguistics Vanguard* 1, no. 1: 131–143.

Enfield, N. J., and Jack Sidnell. 2017a. *The Concept of Action.* Cambridge, UK: Cambridge University Press.

Enfield, N. J., and Jack Sidnell. 2017b. "On the Concept of Action in the Study of Interaction." *Discourse Studies* 19, no. 5: 515–535.

Enfield, N. J., and Jack Sidnell. 2019. "The Normative Nature of Language." In *The Normative Animal? On the Anthropological Significance of Social, Moral, and Linguistic Norms,* edited by Neil Roughley and Kurt Bayertz, 265–278. Oxford: Oxford University Press.

Enfield, N. J., and Tanya Stivers, eds. 2007. *Person Reference in Interaction: Linguistic, Cultural, and Social Perspectives.* Cambridge, UK: Cambridge University Press.

Enfield, N. J., and Charles H. P. Zuckerman. In press. "Moorings: Linguistic Practices and the Tethering of Experience, Status, and Action." *Current Anthropology.*

Enquist, Magnus, and Olof Leimar. 1993. "The Evolution of Cooperation in Mobile Organisms." *Animal Behaviour* 45, no. 4: 747–757.

Errington, James Joseph. 1984. "Self and Self-Conduct among the Javanese Priyayi Elite." *American Ethnologist* 11, no. 2: 275–290.

Errington, James Joseph. 1998. *Shifting Languages: Interaction and Identity in Javanese Indonesia*. Cambridge, UK: Cambridge University Press.

Evans, Nicholas. 2003a. "Context, Culture, and Structuration in the Languages of Australia." *Annual Review of Anthropology* 32: 13–40.

Evans, Nicholas. 2003b. *An Interesting Couple: The Semantic Development of Dyad Morphemes*. Köln: Institut für Sprachwissenschaft der Universität zu Köln.

Evans, Nicholas, Henrik Bergqvist, and Lila San Roque. 2018a. "The Grammar of Engagement I: Framework and Initial Exemplification." *Language and Cognition* 10, no. 1: 110–40.

Evans, Nicholas, Henrik Bergqvist, and Lila San Roque. 2018b. "The Grammar of Engagement II: Typology and Diachrony." *Language and Cognition* 10, no. 1: 141–70.

Evans, Nicholas, and Stephen C. Levinson. 2009. "The Myth of Language Universals: Language Diversity and its Importance for Cognitive Science." *Behavioral and Brain Sciences* 32: 429–492.

Everett, Daniel L. 2005. "Cultural Constraints on Grammar and Cognition in Pirahã: Another Look at the Design Features of Human Language." *Current Anthropology* 46, no. 4: 621–46. https://doi.org/10.1086/431525.

Everett, Daniel L. 2017. *How Language Began: The Story of Humanity's Greatest Invention*. New York: W. W. Norton & Company.

Fausey, Caitlin M., and Lera Boroditsky. 2010. "Subtle Linguistic Cues Influence Perceived Blame and Financial Liability." *Psychonomic Bulletin & Review* 17, no. 5: 644–50.

Finney, Ben R. 1973. *Big Men and Business: Entrepreneurship and Economic Growth in the New Guinea Highlands*. Honolulu: University of Hawai'i Press.

Fleming, Luke. In press. *On Speaking Terms: Avoidance Registers and the Sociolinguistics of Kinship*. Toronto: University of Toronto Press.

Fleming, Luke, and Jack Sidnell. 2020. "The Typology and Social Pragmatics of Interlocutor Reference in Southeast Asia." *The Journal of Asian Linguistic Anthropology* 2, no. 3: 1–20.

Fleming, Luke, and James Slotta. 2018. "The Pragmatics of Kin Address: A Sociolinguistic Universal and Its Semantic Affordance." *Journal of Sociolinguistic* 22, no. 4: 375–405.

Foley, William A., and Robert D. Van Valin Jr. 1984. *Functional Syntax and Universal Grammar*. Cambridge: Cambridge University Press.

Foley, William A. 1986. *The Papuan Languages of New Guinea*. Cambridge: Cambridge University Press.

Fortes, Meyer. 1983. *Rules and the Emergence of Society*. London: Royal Anthropological Institute of Great Britain and Ireland.

Foster, George M. 1966. "Euphemisms and Cultural Sensitivity in Tzintzuntzan." *Anthropological Quarterly* 39, no. 2: 53–59.

Foucault, Michel. 1978. *The History of Sexuality, Vol. 1*. New York: Pantheon Books.

Foucault, Michel. 1980. *Power/Knowledge*. London and New York: Pantheon Books.

Fox, Barbara A., and Sandra A. Thompson. 2010. "Responses to Wh-Questions in English Conversation." *Research on Language and Social Interaction* 43, no. 2: 133–156.

Fraser, Nancy. 1990. "Rethinking the Public Sphere: A Contribution to the Critique of Actually Existing Democracy." *Social Text* 25/26: 58–80.

Gal, Susan, and Katherine Woolard, eds. 2001. *Languages and Publics: The Making of Authority*. London: Routledge.

Gardner, R. 2001. *When Listeners Talk: Response Tokens and Listener Stance*. Amsterdam: John Benjamins.

Garfinkel, Harold. 1967. *Studies in Ethnomethodology*. Englewood Cliffs, NJ: Prentice-Hall.

Geertz, Clifford. 1957. "Ritual and Social Change: A Javanese Example." *American Anthropologist* 59, no. 1: 32–54.

Geertz, Clifford. 1960. *The Religion of Java*. Glencoe, IL: Free Press.

Geertz, Clifford. 1980. *Negara: The Theatre State in Nineteenth Century Bali*. Princeton, NJ: Princeton University Press.

Gilbert, Margaret. 1990. "Walking Together: A Paradigmatic Social Phenomenon." *Midwest Studies in Philosophy* 15, no. 1: 1–14.

Gilbert, Margaret. 1992. *On Social Facts*. London: Routledge.

Givón, Talmy. 1979. *On Understanding Grammar*. New York: Academic Press.

Givón, Talmy. 1984. *Syntax: A Functional-Typological Introduction, Volume I*. Amsterdam ; Philadelphia: John Benjamins.

Givón, Talmy. 1991. *Syntax: A Functional-Typological Introduction, Volume II*. Amsterdam ; Philadelphia: John Benjamins.

Glenn, Phillip. 2003. *Laughter in Interaction*. Cambridge, UK: Cambridge University Press.

Gluckman, Max. 1963. "Gossip and Scandal." *Current Anthropology* 4, no. 3: 307–316.

Goddard, Cliff. 1998. *Semantic Analysis: A Practical Introduction*. Oxford: Oxford University Press.

Goffman, Erving. 1957. "Alienation from Interaction." *Human Relations* 10, no. 1: 47–60.

Goffman, Erving. 1961. "Fun in Games". In *Encounters*, 15–72. Harmondsworth, UK: Penguin.

Goffman, Erving. 1963a. *Behavior in Public Places*. New York: The Free Press.

Goffman, Erving. 1963b. *Stigma: Notes on the Management of Spoiled Identity*. New York: Simon & Schuster.

Goffman, Erving. 1964. "The Neglected Situation." *American Anthropologist* 66, no. 6: 133–136.

Goffman, Erving. 1974. *Frame Analysis: An Essay in the Organization of Experience*. Boston, MA: Northeastern University Press.

Goffman, Erving. 1976. "Replies and Responses." *Language in Society* 5: 257–313.

Goldberg, Adele E. 2006. *Constructions at Work: The Nature of Generalization in Language*. Oxford: Oxford University Press.

Goodwin, Charles. 1979. "The Interactive Construction of a Sentence in Natural Conversation." In *Everyday Language: Studies in Ethnomethodology*, edited by George Psathas, 97–121. New York: Irvington.

Goodwin, Charles. 1980. "Restarts, Pauses, and the Achievement of Mutual Gaze at Turn Beginning." *Sociological Inquiry* 50: 272–302.

Goodwin, Charles. 1981. *Conversational Organization: Interaction between Speakers and Hearers*. New York: Academic Press.

Goodwin, Charles. 1986. "Between and Within: Alternative Treatments of Continuers and Assessments." *Human Studies* 9: 205–217.

Goodwin, Charles. 1994. "Professional Vision." *American Anthropologist* 96, no. 3: 606–633.

Goodwin, Charles. 2000. "Action and Embodiment within Situated Human Interaction." *Journal of Pragmatics* 32, no. 10: 1489–1522.

Goodwin, Charles. 2003. "Pointing as Situated Practice." In *Pointing: Where Language, Culture and Cognition Meet*, edited by Sotaro Kita, 217–241. Mahwah, NJ: Lawrence Erlbaum.

Goodwin, Charles. 2013. "The Co-Operative, Transformative Organization of Human Action and Knowledge." *Journal of Pragmatics* 46, no. 1: 8–23.

Goodwin, Charles, and Marjorie Harness Goodwin. 1987. "Concurrent Operations on Talk: Notes on the Interactive Organization of Assessments." *IPrA Papers in Pragmatics* 1, no. 1: 1–52.

Goodwin, Charles, and Marjorie Harness Goodwin. 2004. "Participation." In *A Companion to Linguistic Anthropology*, edited by Alessandro Duranti, 222–244. Oxford: Blackwell.

Goodwin, Marjorie Harness. 1980. "Processes of Mutual Monitoring Implicated in the Production of Description Sequences." *Sociological Inquiry* 50: 303–317.

Goodwin, Marjorie Harness. 1991. *He-Said-She-Said: Talk as Social Organization among Black Children*. Bloomington: Indiana University Press.

Goodwin, Marjorie Harness, and Charles Goodwin. 1986. "Gesture and Coparticipation in the Activity of Searching for a Word." *Semiotica* 62, no. 1/2: 51–75.

Goody, Esther. 1997. "Social Intelligence and the Emergence of Roles and Rules." *Proceedings of the British Academy* 97: 119–147.

Gramsci, Antonio. 2000. *The Antonio Gramsci Reader: Selected Writings 1916–1935*, edited by David Forgacs. New York: New York University Press.

Greenberg, Joseph Harold. (1966). 2005. *Language Universals: With Special Reference to Feature Hierarchies*. Berlin: Mouton de Gruyter.

Grice, H. Paul. 1957. "Meaning." *Philosophical Review* 66, no. 3: 377–388.

Grice, H. Paul. 1975. "Logic and Conversation." *Syntax and Semantics 3: Speech Acts*, edited by Peter Cole and Jerry L. Morgan, 41–58. New York: Academic Press.

Habermas, Jürgen. 1989. *The Structural Transformation of the Public Sphere: An Inquiry into a Category of Bourgeois Society*. Cambridge, UK: Polity Press.

Hacking, Ian. 1986. "Making Up People." In *Reconstructing Individualism*, edited by Thomas C. Heller, Morton Sosna, and David E. Wellberry, 222–236. Stanford: Stanford University Press.

Hale, Kenneth L. 1966. "Kinship Reflections in Syntax: Some Australian Languages." *Word* 22, no. 1–3: 318–324.

Hale, Kenneth L. 1976. "The Adjoined Relative Clause in Australia." In *Grammatical Categories in Australian Languages*, edited by R. M. W. Dixon, 78–105. Canberra: Australian Institute of Aboriginal Studies.

Halliday, M. A. K., and Ruqaiya Hasan. 1976. *Cohesion in English*. London: Longman.

Hanks, L. 1962. "Merit and Power in the Thai Social Order." *American Anthropologist* 64, no. 6: 1247–1261.

Hanks, William F. 1990. *Referential Practice: Language and Lived Space among the Maya*. Chicago: University of Chicago Press.

Hanks, William F. 2005. "Pierre Bourdieu and the Practices of Language." *Annual Review of Anthropology* 43: 67–83.

Hanks, William F. 2006. "Joint Commitment and Common Ground in a Ritual Event." In *Roots of Human Sociality: Culture, Cognition and Interaction*, edited by N. J. Enfield and Stephen C. Levinson. Oxford: Berg.

Harris, Paul L. 2005. "Conversation, Pretence, and Theory of Mind." In *Why Language Matters for Theory of Mind*, edited by Janet W. Astington and J. A. Baird, Oxford: Oxford University Press. 70–83.

Hart, H. L. A. 1961. *The Concept of Law*. Oxford: Oxford University Press.

Haslam, Nick. 2016. "Concept Creep: Psychology's Expanding Concepts of Harm and Pathology." *Psychological Inquiry* 27, no. 1: 1–17.

Haviland, John B. 1977. *Gossip, Reputation and Knowledge in Zinacantan*. Chicago: University of Chicago Press.

Haviland, John B. 1979. "Guugu Yimidhirr Brother-in-Law Language." *Language in Society* 8, no. 2–3: 365–393.

Haviland, John B. 1991. *Projections, Transpositions and Relativity*. Nijmegen: Cognitive Anthropology Research Group.

Haviland, John B. 1993. "Anchoring, Iconicity, and Orientation in Guugu Yimithirr Pointing Gestures." *Journal of Linguistic Anthropology* 3, no. 1: 3–45.

Haviland, John B. 1996. "Projections, Transpositions, and Relativity." In *Rethinking Linguistic Relativity*, edited by John J. Gumperz and Stephen C. Levinson, 271–323. Cambridge, UK: Cambridge University Press.

Haviland, John B. 1998. "Guugu Yimithirr Cardinal Directions." *Ethos* 26, no. 1: 25–47.

Hawkes, Kristen. 1978. "Big-Men in Binumarien." *Oceania* 48: 161–187.

Hawkins, John A. 2004. *Efficiency and Complexity in Grammars*. Oxford: Oxford University Press.

Hawkins, John A., and Anne Cutler. 1988. "Psycholinguistic Factors in Morphological Asymmetry." *Explaining Language Universals*, edited by John A. Hawkins, 280–317. Oxford: Blackwell.

Hayashi, Makoto, and Kaoru Hayano. 2013. "Proffering Insertable Elements: A Study of Other-Initiated Repair in Japanese." In *Conversational Repair and Human Understanding*, edited by Makoto Hayashi, Geoffery Raymond, and Jack Sidnell, 293–321. Cambridge, UK: Cambridge University Press.

Hayashi, Makoto, Geoffrey Raymond, and Jack Sidnell, eds. 2013. *Conversational Repair and Human Understanding*. Cambridge, UK: Cambridge University Press.

Heritage, John. 1984a. "A Change-of-State Token and Aspects of Its Sequential Placement." In *Structures of Social Action*, edited by J. Maxwell Atkinson and John Heritage, 299–345. Cambridge, UK: Cambridge University Press.

Heritage, John. 1984b. *Garfinkel and Ethnomethodology*. Cambridge, UK: Polity Press.

Heritage, John. 1998. "Oh-Prefaced Responses to Inquiry." *Language in Society* 27: 291–334.

Heritage, John. 2002. "Oh-Prefaced Responses to Assessments: A Method of Modifying Agreement/Disagreement." In *The Language of Turn and Sequence*, edited by Cecilia E. Ford, Barbara A. Fox, and Sandra A. Thompson, 196–224. Oxford: Oxford University Press.

Heritage, John, and Steve Clayman. 2010. *Talk in Action: Interactions, Identities, and Institutions*. Hoboken, NJ: Wiley-Blackwell.

Heritage, John, and Geoffrey Raymond. 2012. "Navigating Epistemic Landscapes: Acquiesence, Agency and Resistance in Responses to Polar Questions." In *Questions: Formal, Functional and Interactional Perspectives*, edited by Jan-Peter de Ruiter, 179–192. Cambridge, UK: Cambridge University Press.

Heritage, John, and Sue Sefi. 1992. "Dilemmas of Advice: Aspects of the Delivery and Reception of Advice in Interactions between Health Visitors and First Time Mothers." In *Talk at Work*, edited by Paul Drew and John Heritage, 359–417. Cambridge, UK: Cambridge University Press.

Herrmann, Esther, Josep Call, Maria Victoria Hernandez-Lloreda, Brian Hare, and Michael Tomasello. 2007. "Humans Have Evolved Specialized Skills of Social Cognition: The Cultural Intelligence Hypothesis." *Science* 317, no. 5843: 1360–1366.

Herrmann, Esther, and Michael Tomasello. 2010. "Ape and Human Cognition: What's the Difference?" *Current Directions in Psychological Science* 19, no. 1: 3–8.

Herzog, George. 1945. "Drum-Signaling in a West African Tribe." *Word* 1, no. 3: 217–238.

Himmelmann, Nikolaus P. 2014. "Asymmetries in the Prosodic Phrasing of Function Words: Another Look at the Suffixing Preference." *Language* 90, no. 4: 927–960.

Hinde, Robert A. 1997. *Relationships: A Dialectical Perspective*. Hove, UK: Psychology Press.

Hockett, Charles F. 1960. "The Origin of Speech." *Scientific American* 203: 88–111.

Horenstein, Sidney. 2009. "The 'Popular Press' Responds to Charles Darwin, *The Origin of Species*, and His Other Works." *Evolution: Education and Outreach* 2: 107–116.

Humboldt, Wilhelm von. [1836] 1988. *On Language: The Diversity of Human Language-Structure and Its Influence on the Mental Development of Mankind*. Cambridge: Cambridge University Press.

Humphrey, Caroline. 1993. "Women, Taboo, and the Suppression of Attention." In *Defining Females: The Nature of Women in Society*, edited by Shirley Arderner, 73–92. Oxford and Providence: Berg.

Humphrey, Nicholas K. 1976. "The Social Function of Intellect." In *Growing Points in Ethology*, edited by P. P. G. Bateson and R. A. Hinde, 303–317. Cambridge, UK: Cambridge University Press.

Hurford, James. 1999. "Functional Innateness: Explaining the Critical Period for Language Acquisition." *Functionalism and Formalism in Linguistics, Vol. 2: Case Studies*, edited by Michael Darnell, Edith A. Moravcsik, Michael Noonan, Frederick J. Newmeyer, and Kathleen Wheatley, 341–363. Amsterdam: John Benjamins.

Hurford, James. 2003. "The Neural Basis of Predicate-Argument Structure." *Behavioral and Brain Sciences* 26, no. 3: 261–282.

Hurford, James. 2007. "The Origins of Meaning: Language in the Light of Evolution." *Functions of Language* 16, no. 2: 298–308.

Hurford, James. 2008. "The Evolution of Human Communication and Language." *Sociobiology of Communication: An Interdisciplinary Perspective*, edited by Patrizia d'Ettoree and David P. Hughes, 249–264. Oxford: Oxford University Press.

Hurford, James. 2012. *The Origins of Grammar, Vol 2*. Oxford: Oxford University Press.

Hutto, Daniel D. 2008. *Folk Psychological Narratives: The Sociocultural Basis of Understanding Reasons*. Cambridge, MA: MIT Press.

Irons, William. 1975. *The Yomut Turkmen: A Study of Social Organization among a Central Asian Turkic-Speaking Population*. Ann Arbor: University of Michigan, Museum of Anthropology.

Jakobson, Roman. 1960. "Closing Statement: Linguistics and Poetics." In *Style in Language*, edited by Thomas Sebeok, 350–377. Cambridge, MA: Technology Press of MIT.

Jakobson, Roman. 1971. "Shifters, Verbal Categories, and the Russian Verb." In *Selected Writings II. Word and Language, 130-147*. The Hague: Mouton

Jakobson, Roman. 1985. "The Fundamental and Specific Characteristics of Human Language." In *Selected Writings VII. Contributions to Comparative Mythology. Studies in Linguistics and Philology, 1972–1982*, 93–97. The Hague: Mouton.

Jakobson, Roman. 1990a. "Linguistics in Relation to Other Sciences." In *On Language*, edited by Linda R. Waugh and Monique Monville-Burston, 451–488. Cambridge, MA: Harvard University Press.

Jakobson, Roman. 1990b. "Shifters and Verbal Categories." In *On Language*, edited by Linda R. Waugh and Monique Monville-Burston, 386–392. Cambridge, MA: Harvard University Press.

Jefferson, Gail. 1972. "Side Sequences." In *Studies in Social Interaction*, edited by David N. Sudnow, 294–338. New York: Free Press.

Jefferson, Gail. 1973. "A Case of Precision Timing in Ordinary Conversation: Overlapped Tag-Positioned Address Terms in Closing Sequences." *Semiotica* 9, no. 1: 47–96.

Jefferson, Gail. 1974. "Error Correction as an Interactional Resource." *Language in Society* 3, no. 2: 181–199.

Jefferson, Gail. 1978. "What's in a 'Nyem'?" *Sociology* 12, no. 1: 135–139.

Jefferson, Gail. 1979. "A Technique for Inviting Laughter and Its Subsequent Acceptance/Declination." In *Everyday Language: Studies in Ethnomethodology*, edited by George Psathas, 79–96. New York: Irvington Publishers.

Jefferson, Gail. 1980. "On 'Trouble-Premonitory' Response to Inquiry." *Sociological Inquiry* 50: 153–185.

Jefferson, Gail. 1984. "On the Organization of Laughter in Talk about Troubles." In *Structures of Social Action: Studies in Conversation Analysis*, edited by J. Maxwell Atkinson and John Heritage, 346–369. Cambridge, UK: Cambridge University Press.

Jefferson, Gail. 1987. "On Exposed and Embedded Correction in Conversation." In *Talk and Social Organization*, edited by Graham Button and John R. E. Lee, 86–100. Clevedon, UK: Multilingual Matters.

Jefferson, Gail, Harvey Sacks, and Emanuel A. Schegloff. 1987. "Notes on Laughter in the Pursuit of Intimacy." In *Talk and Social Organization*, edited by Graham Button and John R. E. Lee, 152–205. Clevedon, UK: Multilingual Matters.

Jolly, Alison. 1966. "Lemur Social Behavior and Primate Intelligence." *Science* 29, no. 153: 501–506.

Keane, Webb. 1997. *Signs of Recognition: Powers and Hazards of Representation in an Indonesian Society*. Berkeley: University of California Press.

Keane, Webb. 2003. "Semiotics and the Social Analysis of Material Things." *Language and Communication* 23: 409–425.

Keesing, Roger M. 1978. *Elota's Story: The Life and Times of a Solomon Islands Big Man*. New York: St Martin's Press.

Keevallik, Leelo. 2010. "Bodily Quoting in Dance Correction." *Research on Language & Social Interaction* 43, no. 4: 401–426.

Kendon, Adam. 1988. "How Gestures Can Become Like Words." In *Cross-Cultural Perspectives in Nonverbal Communication*, edited by Fernando Poyatos, 131–141. Toronto: C. J. Hogrefe.

Kendon, Adam. 1990. *Conducting Interaction: Patterns of Behavior in Focused Encounters*. Cambridge, UK: Cambridge University Press.

Kendon, Adam, and Laura Versante. 2003. "Pointing by Hand in 'Neapolitan.'" In *Pointing*, edited by Sotaro Kita, 109–38. Mahwah, NJ: Lawrence Erlbaum.

Kidwell, Mardi. 2013. "Availability as a Trouble Source in Directive-Response Sequences." In *Conversational Repair and Human Understanding*, edited by Makoto

Hayashi, Geoffrey Raymond, and Jack Sidnell, 234–260. Cambridge, UK: Cambridge University Press.

Kidwell, Mardi, and Zimmerman, Don. 2006. "'Observability' in the Interactions of Very Young Children." *Communication Monographs* 73, no. 1: 1–28.

Kirby, Simon. 2002. "Learning, Bottlenecks and the Evolution of Recursive Syntax." In *Linguistic Evolution through Language Acquisition: Formal and Computational Models*, edited by Ted Briscoe, 173–203. Cambridge, UK: Cambridge University Press.

Kita, Sotaro, ed. 2003. *Pointing: Where Language, Cognition, and Culture Meet.* Mahwah, NJ: Lawrence Erlbaum.

Kita, Sotaro, and James Essegbey. 2001. "Pointing Left in Ghana: How a Taboo on the Use of the Left Hand Influences Gestural Practice." *Gesture* 1, no. 1: 73–94.

Kockelman, Paul. 2005. "The Semiotic Stance." *Semiotica* 157, no. 1–4: 233–304.

Kockelman, Paul. 2006a. "Representations of the World: Memories, Perceptions, Beliefs, Intentions, and Plans." *Semiotica* 162 no. 1–4: 73–125.

Kockelman, Paul. 2006b. "Residence in the World: Affordances, Instruments, Actions, Roles, and Identities." *Semiotica* 162, no. 1–4: 19–71.

Kockelman, Paul. 2007a. "From Status to Contract Revisited: Value, Temporality, Circulation and Subjectivity." *Anthropological Theory* 7, no. 2: 151–176.

Kockelman, Paul. 2007b. "Agency: The Relation between Meaning, Power, and Knowledge." *Current Anthropology* 48, no. 3: 375–401.

Kockelman, Paul. 2010. "Enemies, Parasites, and Noise: How to Take Up Residence in a System without Becoming a Term in It." *Journal of Linguistic Anthropology* 20, no. 2: 406–421.

Kockelman, Paul. 2013. *Agent, Person, Subject, Self: A Theory of Ontology, Interaction, and Infrastructure*. Oxford and New York: Oxford University Press.

Kockelman, Paul. 2017. "Semiotic Agency." In *Distributed Agency*, edited by N. J. Enfield and Paul Kockelman, 25–38. Oxford: Oxford University Press.

Krebs, John R., and Richard Dawkins. 1984. "Animal Signals: Mind-Reading and Manipulation." In *Behavioral Ecology: An Evolutionary Approach*, edited by. John R. Krebs and Nicholas B. Davies, 380–402. Oxford: Blackwell Scientific.

Landaburu, Jon. 2007. "La Modalisation Du Savoir En Langue Andoke (Amazonie Colombienne)." In *L'énonciation Médiatisée II: Le Traitement Épistémologique de l'information; Illustrations Amérindiennes et Caucasiennes*, edited by Zlatka Guentchéva and Jon Landaburu, 23–47. Leuven: Peeters.

Langacker, Ronald W. 1987. *Foundations of Cognitive Grammar: Volume I, Theoretical Prerequisites*. Stanford: Stanford University Press.

Langer, Ellen J., Arthur Blank, and Benzion Chanowitz. 1978. "The Mindlessness of Ostensibly Thoughtful Action: The Role of 'Placebic' Information in Interpersonal Interaction." *Journal of Personality and Social Psychology*, 36, no. 6, 635–642.

LaPolla, Randy J. 2003. "Why Languages Differ: Variation in the Conventionalization of Constraints on Inference." In *Language Variation - Papers on Variation and Change in the Sinosphere and in the Indosphere in Honour of James A. Matisoff*, edited by David Bradley, Randy J. LaPolla, Boyd Michailovsky, and Graham W. Thurgood, 113–44. Canberra: Pacific Linguistics.

Lehtonen, Jaakko, and Kari Sajavaara. 1985. "The Silent Finn." In *Perspectives on Silence*, edited by Deborah Tannen and Muriel Saville-Troike, 193–201. Norwood, NJ: Ablex.

Levelt, Willem J. M. 2013. *A History of Psycholinguistics: The Pre-Chomskyan Era*. Oxford: Oxford University Press.

Levinson, Stephen C. 1983. *Pragmatics*. Cambridge, UK: Cambridge University Press.

Levinson, Stephen C. 2000. *Presumptive Meanings: The Theory of Generalized Conversational Implicature*. Cambridge, MA: MIT Press.

Levinson, Stephen C. 2003. *Space in Language and Cognition: Explorations in Cognitive Diversity*. Cambridge, UK: Cambridge University Press.

Levinson, Stephen C. 2006. "Cognition at the Heart of Human Interaction." *Discourse Studies* 8, no. 1: 85–93.

Levinson, Stephen C. 2014. "Language Evolution." In *The Cambridge Handbook of Linguistic Anthropology*, edited by N. J. Enfield, Paul Kockelman, and Jack Sidnell, 309–324. Cambridge, UK: Cambridge University Press.

Levinson, Stephen C. 2015. "Other-Initiated Repair in Yélî Dnye: Seeing Eye-to-Eye in the Language of Rossel Island." *Open Linguistics* 1, no. 1: 386–410.

Levinson, Stephen C. 2016. "Turn-Taking in Human Communication—Origins and Implications for Language Processing." *Trends in Cognitive Sciences* 20, no. 1: 6–14.

Levinson, Stephen C. 2019. "Interactional Foundations of Language: The Interaction Engine Hypothesis." In *Human Language*, edited by Peter Hagoort, 189–200. Cambridge, MA: MIT Press.

Lewis, David K. 1969. *Convention: A Philosophical Study*. Cambridge, MA: Harvard University Press.

Lindstrom, Lamont. 1981. "Big Man: A Short Terminological History." *American Anthropologist* 83, no. 4: 900–905.

Linton, Ralph. 1936. *The Study of Man*. New York: Appleton-Century.

Liss, Ryan. 2012. "The Abuse of Ambiguity: The Uncertain Status of Omar Khadr under International Law." *The Canadian Yearbook of International Law* 50: 95–161.

Liszkowski, Ulf, Malinda Carpenter, A. Henning, Tricia Striano, and Michael Tomasello. 2004. "Twelve-Month-Olds Point to Share Attention and Interest." *Developmental Science* 7, no. 3: 297–307.

Loftus, Elizabeth F., and John C. Palmer. 1974. "Reconstruction of Automobile Destruction: An Example of the Interaction between Language and Memory." *Journal of Verbal Learning and Verbal Behavior* 13, no. 5: 585–589.

Luong, Hy Van. 1990. *Discursive Practices and Linguistic Meanings: The Vietnamese System of Person Reference*. Amsterdam: John Benjamins Publishing Company.

Lucy, John. 1993. *Reflexive Language: Reported Speech and Metapragmatics*. Cambridge, UK: Cambridge University Press.

Macpherson, C. B. 1962. *The Political Theory of Possessive Individualism: Hobbes to Locke*. Oxford: Clarendon Press.

Malinowski, Bronislaw. 1923. "The Problem of Meaning in Primitive Languages." In *The Meaning of Meaning*, edited by. C. K. Ogden and I. A. Richards, 296–336. London: Routledge and Kegan Paul.

Malinowski, Bronislaw. 1935. *Coral Gardens and their Magic*. London: Allen & Unwin.

Marr, David. 1984. *Vietnamese Tradition on Trial, 1920–1945*. Berkeley: University of California Press.

Martin, Alexander, and Jennifer Culbertson. 2020. "Revisiting the Suffixing Preference: Native Language Affixation Patterns Influence Perception of Sequences." *Psychological Science* 31, no. 9: 1107–1116.

McDaniel, Justin Thomas. 2008. *Gathering Leaves and Lifting Words: Histories of Buddhist Monastic Education in Laos and Thailand*. Seattle: University of Washington Press.

McGregor, William B. 1996. "Dyadic and Polyadic Kin Terms in Gooniyandi." *Anthropological Linguistics* 38, no. 2: 216–247.

McGregor, William B. 2006. "Focal and Optional Ergative Marking in Warrwa (Kimberley, Western Australia)." *Lingua* 116, no. 4: 393–423.

McGregor, William B. 2010. "Optional Ergative Case Marking Systems in a Typological-Semiotic Perspective." *Lingua* 120, no. 7: 1610–1636.

McHale, Shawn Frederick. 2002. *Print and Power: Confucianism, Communism and Buddhism in the Making of Modern Vietnam*. Honolulu: University of Hawai'i Press.

Meissner, Martin, Stuart B. Philpott, and Diana Philpott. 1975a. "A Dictionary of Sawmill Workers' Signs." *Sign Language Studies* 9: 309–348.

Meissner, Martin, Stuart B. Philpott, and Diana Philpott. 1975b. "The Sign Language of Sawmill Workers in British Colombia." *Sign Language Studies* 9: 291–308.

Merlan, Francesca. 1997. "The Mother-in-Law Taboo: Avoidance and Obligation in Aboriginal Australian Society." In *Scholar and Sceptic: Australian Aboriginal Studies in Honour of L. R. Hiatt*, edited by Francesca Merlan, John A. Morton, and Alan Rumsey, 95–122. Canberra: Aboriginal Studies Press.

Merlan, Francesca, and Jeffrey Heath. 1982. "Dyadic Kinship Terms." In *Languages of Kinship in Aboriginal Australia*, edited by Jeffrey Heath, Francesca Merlan, and Alan Rumsey, 107–24. Sydney: University of Sydney.

Milgram, Stanley, Leonard Bickman, and Lawrence Berkowitz. 1969. "Note on the Drawing Power of Crowds of Different Size." *Journal of Personality and Social Psychology* 13, no. 2: 79–82.

Milroy, Lesley, and James Milroy. 1992. "Social Network and Social Class: Toward an Integrated Sociolinguistic Model." *Language in Society* 21, no. 1: 1–26.

Mohr, Susanne. 2015. "Tshaukak'ui—Hunting Signs of the Ts'ixa in Northern Botswana." In *Sign Languages of the World: A Comparative Handbook*, edited by Julie Bakken Jepsen, Goedele de Clerck, Sam Lutalo-Kiingi, and William B. McGregor, 933–954. Berlin: De Gruyter Mouton.

Mowrer, Priscilla. 1970. *Notes on Navajo Silence Behavior*. Unpublished Manuscript. Tuscon: University of New Mexico.

Nelson, Katherine. 2005. "Language Pathways into the Community of Minds." In *Why Language Matters for Theory of Mind*, edited by Janet Wilde Astington and Jodie A. Baird, 26–49. Oxford: Oxford University Press.

Nichols, Johanna. 1984. "Functional Theories of Grammar." *Annual Review of Anthropology* 13: 97–117.

Numbers, Ronald L., and John Stenhouse, eds. 1999. *Disseminating Darwinism: The Role of Place, Race, Religion, and Gender*. Cambridge, UK: Cambridge University Press.

Ochs, Elinor, and Lisa Capps. 1996. "Narrating the Self." *Annual Review of Anthropology* 25: 19–43.

Ochs, Elinor, and Lisa Capps. 2009. *Living Narrative: Creating Lives in Everyday Storytelling*. Cambridge, MA: Harvard University Press.

Ohala, John. 1984. "An Ethological Perspective on Common Cross-Language Utilization of F0 of Voice." *Phonetica* 41: 1–16.

Ohala, John. 1994. "The Frequency Code Underlies the Sound-Symbolic Use of Voice Pitch." In *Sound Symbolism*, edited by Leanne Hinton, Johanna Nichola, and John J. Ohala, 325–347. Cambridge, UK: Cambridge University Press.

Origgi, Gloria. 2018. *Reputation: What It Is and Why It Matters*. Princeton, NJ: Princeton University Press.

Parmentier, Richard J. 1994. *Signs in Society: Studies in Semiotic Anthropology*. Bloomington: Indiana University Press.

Peirce, Charles Sanders. (1904) 1998. "What Is a Sign?" In *The Essential Peirce: Selected Philosophical Writings, vol. 2: 1893–1913*, edited by Peirce Edition Project, 4–10. Bloomington: Indiana University Press.

Peirce, Charles Sanders. 1933. *The Collected Papers of Charles Sanders Peirce, Vol. III: Exact Logic*, edited by Charles Hartshorne and Paul Weiss. Cambridge, MA: Harvard University Press.

Peirce, Charles Sanders. 1955. *The Philosophical Writings of Peirce*, edited by Justus Buchler. New York: Dover Publications.

Peirce, Charles Sanders. 1992. *The Essential Peirce, Selected Philosophical Readings Vol. 1*, edited by Nathan Houser and Christian J. W. Kloesel. Bloomington: Indiana University Press.

Peräkylä, Anssi. 2012. "CA in Psychotherapy." In *The Handbook of Conversation Analysis*, edited by Jack Sidnell and Tanya Stivers, 551–574. Oxford: Blackwell.

Pomerantz, Anita. 1978. "Compliment Responses: Notes on the Co-Operation of Multiple Constraints." In *Studies in the Organization of Conversational Interaction*, edited by Jim Schenkein, 79–112. New York: Academic Press.

Pomerantz, Anita. 1984. "Agreeing and Disagreeing with Assessments: Some Features of Preferred/Dispreferred Turn Shapes." In *Structures of Social Action: Studies in Conversation Analysis*, edited by J. Maxwell Atkinson and John Heritage, 57–101. Cambridge, UK: Cambridge University Press.

Popper, Karl. [1974] 2005. *Unended Quest: An Intellectual Autobiography*. London and New York: Routledge.

Putnam, Hilary. 1975. "The Meaning of 'Meaning.'" In *Language, Mind, and Knowledge*, edited by Keith Gunderson, 131–193. Minneapolis: University of Minnesota Press.

Rae, John. 2008. "Lexical Substitution as a Therapeutic Resource." In *Conversation Analysis and Psychotherapy*, edited by Anssi Peräkylä, Charles Antaki, Sanna Vehviläinen, and Ivan Leudar, 62–79. Cambridge, UK: Cambridge University Press.

Raymond, Chase Wesley. 2019. "Intersubjectivity, Normativity, and Grammar." *Social Psychology Quarterly* 82, no. 2: 182–204.

Raymond, Geoffrey. 2003. "Grammar and Social Organization: Yes/No Interrogatives and the Structure of Responding." *American Sociological Review* 68: 939–967.

Raymond, Geoffrey, and Jack Sidnell. 2014. "Conversation Analysis." In *The Discourse Reader*, (3rd ed.), edited by Adam Jaworski and Nikolas Coupland, 249–263. London: Routledge.

Raymond, Geoffrey, and Jack Sidnell. 2019. "Interactions at the Boundaries of a World Known in Common: Initiating Repair with 'What Do You Mean?'" *Research on Language and Social Interaction* 52, no. 2: 117–192.

Reisman, Karl. 1974. "Contrapuntal Conversations in an Antiguan Village." In *Explorations in the Ethnography of Speaking*, edited by Richard Bauman and Joel Sherzer, 110–124. Cambridge, UK: Cambridge University Press.

Roberts, Sean G., and Stephen C. Levinson. 2017. "Conversation, Cognition, and Cultural Evolution: A Model of the Cultural Evolution of Word Order through Pressures Imposed from Turn Taking in Conversation." *Interaction Studies* 18, no. 3: 402–442.

Roberts, Sean G., Francisco Torreira, and Stephen C. Levinson. 2015. "The Effects of Processing and Sequence Organization on the Timing of Turn Taking: A Corpus Study." *Frontiers in Psychology* 6: 509–524.

Robinson, Jeffrey D. 2006. "Managing Trouble Responsibility and Relationships during Conversational Repair." *Communication Monographs* 73, no. 2: 137–161.

Robinson, Jeffrey D. 2014. "What 'What?' Tells Us about How Conversationalists Manage Intersubjectivity." *Research on Language and Social Interaction* 47, no. 2: 109–129.

Rossano, Federico. 2013. "Sequence Organization and Timing of Bonobo Mother-Infant Interactions." *Interaction Studies* 14, no. 2: 160–189.

Rossi, Giovanni. 2014. "When Do People Not Use Language to Make Requests?" In *Requesting in Social Interaction*, edited by Paul Drew & Elizabeth Couper-Kuhlen, 303–334. Amsterdam: John Benjamins.

Rossi, Giovanni. 2020. "Other-Repetition in Conversation across Languages: Bringing Prosody into Pragmatic Typology." *Language in Society* 49, no. 4: 495–520.

Rumsey, Alan. 2010. "'Optional' Ergativity and the Framing of Reported Speech." *Lingua* 120, no. 7: 1652–1676.

Rumsey, Alan. 2021. "Reported Speech and Represented Speech." In *International Encyclopedia of Linguistic Anthropology*, edited by James Stanlaw, 1683–1694. Hoboken, NJ: Wiley.

Rumsey, Alan, Lauren W. Reed, and Francesca Merlan. 2020. "Ku Waru Clause Chaining and the Acquisition of Complex Syntax." *Frontiers in Communication* 5, 19: 1-30.

Ryle, Gilbert. 1949. *The Concept of Mind*. London: Hutchinson.

Sacks, Harvey. 1972. "On the Analyzability of Stories by Children." In *Directions in Sociolinguistics: The Ethnography of Communication*, edited by John J. Gumperz and Dell Hymes, 325–345. New York: Rinehart & Winston.

Sacks, Harvey. [1973] 1987. "On the Preferences for Agreement and Contiguity in Sequences in Conversation." In *Talk and Social Organisation*, edited by Graham Button and John R. E. Lee, 54–69. Clevedon, UK: Multilingual Matters.

Sacks, Harvey. 1975. "Everyone Has to Lie." In *Sociocultural Dimensions of Language Use*, edited by Mary Sanches and Ben G. Blount, 57–80. New York: Academic Press.

Sacks, Harvey. 1979. "Hotrodder: A Revolutionary Category." In *Everyday Language: Studies in Ethnomethodology*, edited by George Psathas, 7–14. New York: Irvington Publishers.

Sacks, Harvey, and Emanuel A. Schegloff. [1979] 2007. "Two Preferences in the Organization of Reference to Persons in Conversation and Their Interaction." In *Person Reference in Interaction: Linguistic, Cultural, and Social Perspectives*, edited by N. J. Enfield and Tanya Stivers. Cambridge, UK: Cambridge University Press.

Sacks, Harvey. 1987. "On the Preferences for Agreement and Contiguity in Sequences in Conversation." In *Talk and Social Organisation*, edited by Graham Button and John R. E. Lee, 54–69. Clevedon, UK: Multilingual Matters.

Sacks, Harvey. 2004. "An Initial Characterization of the Organization of Speaker Turn-Taking in Conversation." In *Conversation Analysis: Studies from the First Generation*, edited by Gene Lerner, 35–42. Amsterdam and Philadelphia: John Benjamins.

Sacks, Harvey, Emanuel A. Schegloff, and Gail Jefferson. 1974. "A Simple Systematic for the Organisation of Turn Taking in Conversation." *Language* 50, no. 4: 696–735.

Sahlins, Marshall. 1963. "Poor Man, Rich Man, Big Man, Chief: Political Types in Polynesia and Melanesia." *Comparative Studies in Society and History* 5: 285–303.

San Roque, Lila, Simeon Floyd, and Elisabeth Norcliffe. 2017. "Evidentiality and Interrogativity." *Lingua* 186/187: 120–143.

San Roque, Lila, Simeon Floyd, and Elisabeth Norcliffe. 2018. "Egophoricity: An Introduction." In *Egophoricity*, edited by Simeon Floyd, Elisabeth Norcliffe, and Lila San Roque, 118: 1–78. Amsterdam: John Benjamins.

Sapir, Edward. 1913. "A Note on Reciprocal Terms of Relationship." *American Anthropologist* 15: 132–138.

Sapir, Edward. 1921. *Language: An Introduction to the Study of Speech*. Orlando/San Diego/New York/London: Harcourt Brace Jovanovich.

Sarvasy, Hannah S. 2020. "The Acquisition of Clause Chaining in Nungon." *Frontiers in Psychology* 11: 1456.

Scheflen, Albert E. 1976. *Human Territories: How We Behave in Space-Time*. Englewood Cliffs, NJ: Prentice-Hall.

Schegloff, Emanuel A. 1967. *The First Five Seconds: The Order of Conversational Opening*. PhD diss., University of California.

Schegloff, Emanuel A. 1968. "Sequencing in Conversational Openings." *American Anthropologist* 70: 1075–1095.

Schegloff, Emanuel A. 1972. "Notes on a Conversational Practice: Formulating Place." In *Studies in Social Interaction*, edited by David Sudnow, 75–119. New York: Free Press.

Schegloff, Emanuel A. 1982. "Discourse as an Interactional Achievement: Some Uses of 'Uh Huh' and Other Things That Come between Sentences." In *Georgetown University Round Table on Languages and Linguistics 1981: Analyzing Discourse: Text and Talk*, edited by Deborah Tannen 71–93. Washington, DC: Georgetown University Press.

Schegloff, Emanuel A. 1987. "Some Sources of Misunderstanding in Talk-in-Interaction." *Linguistics* 25, 201-218.

Schegloff, Emanuel A. 1988a. "Description in the Social Sciences I: Talk-in-Interaction." *IPrA Papers in Pragmatics*, 2, no. 1–2: 1–24.

Schegloff, Emanuel A. 1988b. "Goffman and the Analysis of Conversation." In *Erving Goffman: Exploring the Interaction Order*, edited by Paul Drew and Anthony Wootton, 89–135. Cambridge, UK: Polity Press.

Schegloff, Emanuel A. 1988c. "Presequences and Indirection: Applying Speech Act Theory to Ordinary Conversation." *Journal of Pragmatics* 12: 55–62.

Schegloff, Emanuel A. 1989. "Reflections on Language, Development, and the Interactional Character of Talk-in-Interaction." In *Interaction in Human Development*, edited by Marc H. Bornstein and Jerome S. Bruner, 139–153. Hillsdale, NJ: Lawrence Erlbaum.

Schegloff, Emanuel A. 1991. "Conversation Analysis and Socially Shared Vognition." In *Perspectives on Socially Shared Cognition*, edited by Lauren B. Resnick, John M. Levine, and Stephanie D. Teasley, 150–171. Washington, DC: American Psychological Association.

Schegloff, Emanuel A. 1992. "Repair after Next Turn: The Last Structurally Provided for Place for the Defense of Intersubjectivity in Conversation." *American Journal of Sociology* 95: 1295–1345.

Schegloff, Emanuel A. 1996. "Confirming Allusions: Toward an Empirical Account of Action" *American Journal of Sociology* 104: 161–216.

Schegloff, Emanuel A. 1997. "Practices and Actions: Boundary Cases of Other-Initiated Repair." *Discourse Processes* 23: 499–545.

Schegloff, Emanuel A. 2000a. "On Granularity." *Annual Review of Sociology* 26: 715–720.

Schegloff, Emanuel A. 2000b. "Overlapping Talk and the Organization of Turn-Taking for Conversation." *Language in Society* 29: 1–63.

Schegloff, Emanuel A. 2006. "Interaction: The Infrastructure for Social Institutions, the Natural Ecological Niche for Language, and the Arena in which Culture Is Enacted." In *Roots of Human Sociality*, edited by N. J. Enfield and Stephen C. Levinson, 70–96. London: Berg.

Schegloff, Emanuel A. 2007a. *Sequence Organization in Interaction: A Primer in Conversation Analysis*. Cambridge, UK: Cambridge University Press.

Schegloff, Emanuel A. 2007b. "A Tutorial on Membership Categorization." *Journal of Pragmatics*, 39, no. 3: 462–482.

Schegloff, Emanuel A. 2013. "Ten Operations in Self-Initiated, Same-Turn Repair." In *Conversational Repair and Human Understanding*, edited by Makoto Hayashi, Geoffrey Raymond, and Jack Sidnell, 41–70. Cambridge, UK: Cambridge University Press.

Schegloff, Emanuel A., Gail Jefferson, and Harvey Sacks. 1977. "The Preference for Self-Correction in the Organization of Repair in Conversation." *Language* 53: 361–382.

Schegloff, Emanuel A., and Harvey Sacks. 1973. "Opening Up Closings." *Semiotica* 8: 289–327.

Schelling, Thomas. 1960. *The Strategy of Conflict*. Cambridge, MA: Harvard University Press.

Scott-Phillips, Thom. 2015. *Speaking Our Minds: Why Human Communication Is Different and How Language Evolved to Make It Special*. New York: Palgrave-Macmillan.

Searle, John R. 1969. *Speech Acts: An Essay in the Philosophy of Language*. Cambridge: Cambridge University Press.

Searle, John R. 1995. *The Construction of Social Reality*. New York: Free Press.

Searle, John R. 2010. *The Making of the Social World: The Structure of Human Civilization*. Oxford: Oxford University Press.

Seifart, Frank, Julien Meyer, Sven Grawunder, and Laure Dentel. 2018. "Reducing Language to Rhythm: Amazonian Bora Drummed Language Exploits Speech Rhythm for Long-Distance Communication." *Royal Society Open Science* 5, no. 4.

Shephard, Michelle. 2008. *Guantanamo's Child: The Untold Story of Omar Khadr*. New York: John Wiley and Sons.

Sherzer, Joel. 1973. "Verbal and Nonverbal Deixis: The Pointed Lip Gesture among the San Blas Cuna." *Language in Society* 2: 117–31.

Sidnell, Jack. 2001. "Conversational Turn-Taking in a Caribbean English Creole." *Journal of Pragmatics* 33: 1263–1290.

Sidnell, Jack. 2003. "An Ethnographic Consideration of Rule-Following." *Journal of the Royal Anthropological Society* 9, no. 3: 429–445.

Sidnell, Jack. 2005. *Talk and Practical Epistemology: The Social Life of Knowledge in a Caribbean Community*. Amsterdam and Philadelphia: John Benjamins.

Sidnell, Jack. 2007. "'Look'-Prefaced Turns in First and Second Position: Launching, Interceding and Redirecting Action." *Discourse Studies* 9: 387–408.

Sidnell, Jack. 2009a. "Comparative Perspectives in Conversation Analysis." In *Conversation Analysis: Comparative Perspectives*, edited by Jack Sidnell, 3–33. Cambridge, UK: Cambridge University Press.

Sidnell, Jack, ed. 2009b. *Conversation Analysis: Comparative Perspectives*. Cambridge, UK: Cambridge University Press.

Sidnell, Jack. 2009c. "Participation." In *The Pragmatics of Interaction*, edited by Sigurd D'Hondt, Jan-Ola Östman, and Jef Verschuren, 125–156. Amsterdam: John Benjamins.

Sidnell, Jack. 2010. *Conversation Analysis: An Introduction*. Chichester: Wiley-Blackwell.

Sidnell, Jack. 2011. "'D'You Understand That Honey?': Gender and Participation in Conversation." In *Conversation and Gender*, edited by Susan A. Speer and Elizabeth Stokoe, 183–209. Cambridge, UK: Cambridge University Press.

Sidnell, Jack. 2014. "The Architecture of Intersubjectivity Revisited." In *The Cambridge Handbook of Linguistic Anthropology*, edited by N. J. Enfield, Paul Kockelman, and Jack Sidnell, 364–399. Cambridge, UK: Cambridge University Press.

Sidnell, Jack. 2017. "Action in Interaction Is Conduct under a Description." *Language in Society* 46: 313–337.

Sidnell, Jack. 2021. "Generic Reference and Social Ontology in Vietnamese Conversation." *Language in Society* 50, no. 4: 533–555.

Sidnell, Jack. 2022. "'Respect Those above, Yield to Those Below': Civility and Social Hierarchy in Vietnamese Interlocutor Reference." In *Signs of Deference, Signs of Demeanour: Interlocutor Reference and Self-Other Relaitons across Southeast Asian Speech Communities*, edited by Dwi Noverini Djenar, and Jack Sidnell. Singapore: National University of Singapore Press.

Sidnell, Jack, and Rebecca Barnes. 2013. "Alternative, Subsequent Descriptions." In *Conversational Repair and Human Understanding*, edited by Makoto Hayashi, Geoffrey Raymond, and Jack Sidnell, 322–342. Cambridge, UK: Cambridge University Press.

Sidnell, Jack, and N. J. Enfield. 2012. "Language Diversity and Social Action: A Third Locus of Linguistic Relativity." *Current Anthropology* 53, no. 3: 302–333.

Sidnell, Jack, and N. J. Enfield. 2017. "How to Distinguish a Wink from a Twitch." *Hau: Journal of Ethnographic Theory* 7, no. 2: 457–465.

Sidnell, Jack, Thùy An Trần, and Thị Thanh Hu'o'ng Vũ. 2020. "On the Division of Intersubjective Labor in Interaction: A Preliminary Study of Other-Initiated Repair in Vietnamese Conversation." In *Studies in the Anthropology of Language in Mainland Southeast Asia. Journal of the Southeast Asian Linguistics Society*, Special Publication no. 6, edited by N. J. Enfield, Jack Sidnell, and Charles H. P. Zuckerman, 65–84. Honolulu: University of Hawai'i Press.

Siegel, James T. 1986. *Solo in the New Order: Language and Hierarchy in an Indonesian City*. Princeton, NJ: Princeton University Press.

Sillitoe, Paul. 1978. "Big Men and War in New Guinea." *Man* 13: 252–271.

Silverstein, Michael. 1976. "Shifters, Linguistics Categories, and Cultural Description." In *Meaning in Anthropology*, edited by Keith H. Basso and Henry A. Selby, 11–55. Albuquerque: University of New Mexico Press.

Silverstein, Michael. 1979. "Language Structure and Linguistic Ideology." In *The Elements: A Parasession on Linguistic Units and Levels*, edited by Paul Clyne, William F. Hanks, and Carol L. Hofbauer, 193–247. Chicago: Chicago Linguistic Society, University of Chicago.

Silverstein, Michael. 1985. "Language and the Culture of Gender: At the Intersection of Structure, Usage, and Ideology." In *Semiotic Mediation: Sociocultural and Psychological Perspectives*, edited by Elizabeth Mertz and Richard Parmentier, 219–259. Orlando FL: Academic Press.

Silverstein, Michael. 1987. "The Three Faces of Function." In *Social and Functional Approaches to Language and Thought*, edited by Maya Hickmann, 17–38. New York: Academic Press.

Silverstein, Michael. 1993. "Metapragmatic Discourse and Metapragmatic Function." In *Reflexive Language: Reported Speech and Metapragmatics*, edited by John Lucy, 33–58. Cambridge, UK: Cambridge University Press.

Slotta, James. 2021. "Pragmatics." In *The International Encyclopedia of Linguistic Anthropology*, edited by James Stanlaw, 1602–1618. Hoboken, NJ: John Wiley and Sons.

Smith, Kenny, Henry Brighton, and Simon Kirby. 2003. "Complex Systems in Language Evolution: The Cultural Emergence of Compositional Structure." *Advances in Complex Systems* 6: 537–558.

Sperber, Dan. 1985. "Anthropology and Psychology: Towards an Epidemiology of Representations." *Man* 20: 73–89.

Sperber, Dan, and Nicolas Baumard. 2012. "Moral Reputation: An Evolutionary and Cognitive Perspective." *Mind and Language* 27, no. 5: 495–518.

Sperber, Dan, and D. Wilson. 1986. *Relevance: Communication and Cognition*. Oxford: Basil Blackwell.

Stalnaker, Robert C. 1999. *Context and Content: Essays on Intentionality in Speech and Thought*. Oxford: Oxford University Press.

Stalnaker, Robert C. 2002. "Common Ground." *Linguistics and Philosophy* 25: 701–21.

Stasch, Rupert. 2009. *Society of Others: Kinship and Mourning in a West Papuan Place*. Berkeley: University of California Press.

Stivers, Tanya. 2005. "Modified Repeats: One Method for Asserting Primary Rights from Second Position." *Research on Language and Social Interaction* 38: 131–158.

Stivers, Tanya. 2007. "Alternative Recognitionals in Person Reference." In *Person Reference in Interaction*, edited by N. J. Enfield, 73–96. Cambridge, UK: Cambridge University Press.

Stivers, Tanya. 2008. "Stance, Alignment, and Affiliation during Storytelling: When Nodding Is a Token of Affiliation." *Research on Language and Social Interaction* 41, no. 1: 31–57.

Stivers, Tanya, N. J. Enfield, Penelope Brown, Christina Englert, Makoto Hayashi, Trine Heinemann, Gertie Hoymann, Federico Rossano, Jan-Peter de Ruiter, Kyung-Eun Yoon, and Stephen C. Levinson. 2009. "Universals and Cultural Variation in Turn-Taking in Conversation." *Proceedings of the National Academy of Sciences of the United States of America* 106, no. 26: 10587–10592.

Stivers, Tanya, and Makoto Hayashi. 2010. "Transformative Answers: One Way to Resist a Question's Constraints." *Language in Society*, 39, no. 1: 1–25.

Streeck, Jürgen. 1993. "Gesture as Communication I: Its Coordination with Gaze and Speech." *Communication Monographs* 60: 275–299.

Szczepek Reed, Beatrice, and Geoffery Raymond, eds. 2013. *Units of Talk—Units of Action*. Amsterdam: John Benjamins Publishing.

Taylor, Charles. 1985. *Philosophical Papers Vol. 1: Human Agency and Language*. Cambridge, UK: Cambridge University Press.

Taylor, Talbot J. 2000. "Language Constructing Language: The Implications of Reflexivity for Linguistic Theory." *Language Sciences* 22, no. 4: 482–499.

Terasaki, Alene Kiku. (1976) 2004. "Pre-Announcement Sequences in Conversation." In *Conversation Analysis: Studies from the First Generation*, edited by Gene Lerner, 171–223. Amsterdam: John Benjamins.

Thompson, Laurence C. 1987. *A Vietnamese Grammar*. Seattle: University of Washington Press.

Thompson, Sandra A., Barbara A. Fox, and Elizabeth Couper-Kuhlen. 2015. *Grammar in Everyday Talk: Building Responsive Actions*. Cambridge, UK: Cambridge University Press.

Tinbergen, N. 1963. "On Aims and Methods of Ethology." *Zeitschrift für Tierpsychologie* 20: 410–433.

Tomasello, Michael. 1999. *The Cultural Origins of Human Cognition*. Cambridge, MA: Harvard University Press.

Tomasello, Michael. 2003. *Constructing a Language: A Usage-Based Theory of Language Acquisition*. Cambridge, MA: Harvard University Press.

Tomasello, Michael. 2008. *The Jean Nicod Lectures: Origins of Human Communication*. Cambridge, MA: MIT Press.

Tomasello, Michael. 2010. *The Origins of Human Communication*. Cambridge, MA: MIT Press.

Tomasello, Michael, Malinda Carpenter, Josep Call, Tanya Behne, and Heinrike Moll. 2005. "Understanding and Sharing Intentions: The Origins of Cultural Cognition." *Behavioral and Brain Sciences* 28: 675–735.

Tomasello, Michael, and Esther Herrmann. 2010. "Ape and Human Cognition: What's the Difference?" *Current Directions in Psychological Science* 19, no. 1: 3–8.

Trudgill, Peter. 2010. *Investigations in Sociohistorical Linguistics*. Cambridge, UK: Cambridge University Press.

Trudgill, Peter. 2011. *Sociolinguistic Typology: Social Determinants of Linguistic Complexity*. Oxford: Oxford University Press.

Tuomela, Raimo. 2002. *The Philosophy of Social Practices: A Collective Acceptance View*. Cambridge, UK: Cambridge University Press.

Tuomela, Raimo. 2007. *The Philosophy of Sociality: The Shared Point of View*. Oxford: Oxford University Press.

Van Valin, R. D., and R. J. LaPolla. 1997. *Syntax: Structure, Meaning, and Function*. Cambridge: Cambridge University Press.

Vološinov, Valentin Nikolaevich. (1929) 1973. *Marxism and the Philosophy of Language*. London: Seminar Press.

Waismann, Friedrich. 1945. "Verifiability." *Proceedings of the Aristotelian Society* 19: 119–50.

Warner, Michael. 2002. *Publics and Counterpublics*. Cambridge, MA: Zone Books.

Weber, Max. 1947. *The Theory of Social and Economic Organization*. New York: Oxford University Press.

Wierzbicka, Anna. 1985. *Lexicography and Conceptual Analysis*. Ann Arbor, MI: Karoma.

Wierzbicka, Anna. 1988. *The Semantics of Grammar*. Amsterdam: John Benjamins.

Wierzbicka, Anna. 1996. *Semantics: Primes and Universals*. Oxford: Oxford University Press.

Wilkins, David P. 2003. "Why Pointing with the Index Finger Is Not a Universal (in Socio-Cultural and Semiotic Terms)." In *Pointing*, edited by Sotaro Kita, 171–216. Mahwah, NJ: Lawrence Erlbaum Associates.

Wilkinson, Sue, and Celia Kitzinger. 2006. "Surprise as an Interactional Achievement: Reaction Tokens in Conversation." *Social Psychology Quarterly* 69: 150–182.

Williams, Bernard. [1985] 2011. *Ethics and the Limits of Philosophy* (Routledge Classics Edition). London: Routledge.

Wittgenstein, Ludwig. 1953. *Philosophical Investigations*. Englewood Cliffs, NJ: Prentice Hall.

Wootton, Anthony. 1997. *Interaction and the Development of Mind*. Cambridge, UK: Cambridge University Press.

Zipf, George K. 1935. *The Psycho-Biology of Language: An Introduction to Dynamic Philology*. Cambridge, MA: MIT Press.

Zipf, George K. 1949. *Human Behavior and the Principle of Least Effort*. Cambridge, MA: Addison-Wesley.

Zlatev, Jordan, Timothy P. Racine, Christopher Sinha, and Esa Itkonen. 2008. "Intersubjectivity: What Makes Us Human?" In *Converging Evidence in Language and Communication Research (CELCR): Vol. 12. The Shared Mind: Perspectives on Intersubjectivity*, edited by Jordan Zlatev, Timothy P. Racine, Christopher Sinha, and Esa Itkonen, 1–14. Amsterdam: John Benjamins.

Zuckerman, Charles H. P. 2016. "Phatic Violence? Gambling and the Arts of Distraction in Laos." *Journal of Linguistic Anthropology* 26, no. 3: 294–314.

Zuckerman, Charles H. P. 2021. "Figure Composition." *Signs and Society* 9, no. 3: 263–299.

Zuckerman, Charles H. P., and N. J. Enfield. 2020. "Heavy Sound Light Sound: A Nam Noi Metalinguistic Trope." In *Studies in the Anthropology of Language in Mainland Southeast Asia*, edited by N. J. Enfield, Jack Sidnell, and Charles H. P. Zuckerman, 85–92. Honolulu: University of Hawai'i Press.

Zuckerman, Charles H. P., and Enfield, N. J. 2022. "The Unbearable Heaviness of Being Kri: House Construction and Ethnolinguistic Transformation in Upland Laos." *Journal of The Royal Anthropological Institute* 28, no. 1: 178–203.

Index